W9-BLT-466

GRAVE SECRETS OF DINOSAURS

GRAVE SECRETS OF DINOSAURS

SOFT TISSUES AND HARD SCIENCE

PHILLIP MANNING

NATIONAL GEOGRAPHIC
WASHINGTON, D.C.

Published by the National Geographic Society

ISBN 978-1-4262-0219-3

Library of Congress Cataloging-in-Publication Data

Manning, Phillip Lars, 1967-
Grave secrets of dinosaurs : soft tissues and hard science / by Phillip Lars Manning.
p. cm.
ISBN 978-1-4262-0219-3 (hardcover : alk. paper)
1. Dinosaurs. 2. Mummified animals. 3. Paleontology. I. Title.
QE861.4.M34 2008
567.9--dc22

2007042938

Founded in 1888, the National Geographic Society is one of the largest nonprofit scientific and educational organizations in the world. It reaches more than 285 million people worldwide each month through its official journal, NATIONAL GEOGRAPHIC, and its four other magazines; the National Geographic Channel; television documentaries; radio programs; films; books; videos and DVDs; maps; and interactive media. National Geographic has funded more than 8,000 scientific research projects and supports an education program combating geographic illiteracy.

For more information, please call 1-800-NGS LINE (647-5463)
or write to the following address:

National Geographic Society
1145 17th Street N.W.
Washington, D.C. 20036-4688 U.S.A.

Visit us online at www.nationalgeographic.com/books

For information about special discounts for bulk purchases, please contact
National Geographic Books Special Sales: ngspecsales@ngs.org

For rights or permissions inquiries, please contact National Geographic Books
Subsidiary Rights: ngbookrights@ngs.org

Interior design: Cameron Zotter and Nicole DiPatrizio

Dedicated to prairie dogs everywhere!

CONTENTS

INTRODUCTION

LATE IN 1999 A YOUNG PALEONTOLOGIST, Tyler Lyson, was heading home after a day of fossil prospecting on his uncle's land in the Hell Creek Formation of North Dakota. Water was running low, so it was time to head back home. As Tyler entered a small valley, bounded by a long, low butte on one side and a high domelike butte to the other, he noticed the telltale shapes that indicated a fossil bone.

Tyler had lived in North Dakota his entire life. Much of his childhood had been spent exploring the Badlands that surrounded and underpinned his hometown of Marmarth. From an early age he had shown a keen interest in the fossil remains littering the landscape where he and his brother often hunted for more modern prey. On his trips he was used to meeting a steady stream of academics that also hunted the Badlands—but for an extinct variety of prey. Tyler was a sharp, intelligent young man who was hungry for knowledge and keen to learn. Most children have a period in their life when dinosaurs are a passion, but few can fuel their passion with vast quantities of fossil remains on their doorstep. Many visiting academics recall the incredible fossils that the young hunter had already plucked from the Badlands.

As Tyler crouched down to look at the find, he knew instantly he had found part of a dinosaur tail. This particular tale about a tail has a twist, given that these bones had weathered out of the stone where they had been lodged for more than 65 million years. Such exposure sometimes means there's more where that came from. A combination of searing summer heat, winter snows, and heavy rain not only sculpted the Badlands, but also has revealed many of the fossils entombed within the sediments.

Gravity is a useful tool for paleontologists in the field. Fossil bones do not roll uphill. So Tyler naturally examined the rock surface immediately above where the exposed tail vertebrae lay. His eye soon spotted the telltale sign of where the vertebrae had fallen from the rock. A patch of bone was still clinging to the rock face, waiting for the next heavy rain storm to reunite it with the bones at the bottom of the slope.

Tyler scrambled up the slope. Just like a layer cake, the sediments of a former floodplain in the Late Cretaceous Period had slowly built up over many years, creating the layers of Hell Creek Formation. As Tyler carefully dug around the tail vertebrae, he saw more buried beneath the surface.

At that sight he hesitated. To start opening a new fossil site at the end of a field season is not a good idea. The problem of someone else discovering the site was not one of Tyler's worries, since he was out in the back of beyond. His main concern was the coming winter. The heating, freezing, and washing with rainwater takes its toll on fossils that are exposed at the surface. Instead Tyler took out his notebook and GPS receiver to pinpoint the fossil's position. In the same way a car's satellite navigation system can nail down where you are on Earth, these systems help us record the positions of paleontological sites so that we can relocate them with ease. Taking GPS coordinates for the site, he noted the location in his field notebook and carefully covered the bones. Tyler then packaged and labeled the bones that had already fallen from the site and brought his hand-sized parcel back to Marmarth. The shape of the bones told Tyler that he had found a dinosaur called a hadrosaur—but how much, he wondered, was buried back where he'd been?

The answer was, incredibly: all of it. For the first time in history a dinosaur with all of its skin, except around a chest wound, would be unearthed. As readers familiar with seeing dinosaurs in museums are aware, the fossil record rarely preserves entire animals. Usually their deaths have left only tough fossil remains of shell, exoskeleton, and bone. From these remains paleontologists are tasked with charting the depths of the history of life on Earth. Yet in rare cases the creatures' soft tissue has been preserved in stone. Such remarkable fossils enable immense advances in our understanding of long-vanished lives and forgotten worlds.

Dinosaur bones have fascinated human beings for thousands of years. Once upon a time such giant bones were reconstructed into the leviathans of myth and legend. They were not formally classified until the 19th century. Given that reconstructions are based upon the evidence available at a particular point in history, the restorations of dinosaurs have evolved through time. Until the mid-20th century dinosaurs were regarded as "failures" while we considered ourselves the ultimate expression of the success of life on Earth. The word "dinosaur" was at that time used to indicate cumbersome, outdated, doomed to fail, or unsuited to environment. Only since the 1970s have dinosaurs been "rehabilitated," regaining their power to awe a popular audience.

Since the naming of the Hell Creek Formation 100 years ago, the fossil remains from the desolate Badlands have inspired generations of paleontologists to hunt for the elusive treasures of past life. These vast deposits contain some of the last fauna and flora from the age of the dinosaurs. Like many paleontologists before me, I, too, have been drawn to this fertile ground of *T. rex* and its kin. Five years would pass before Tyler returned to the hadrosaur site. Only then did he realize the unique potential of this fossil. This is when our paths first crossed and when we started on our exciting journey of discovery.

I am currently employed as a lecturer in paleontology in the School of Earth, Atmospheric, and Environmental Sciences at the University of Manchester in England and also as a Research Fellow at the Manchester Museum. For more than 20 years I have had the good fortune to concentrate much of my efforts on dinosaurs. The fossilized bones of the hadrosaur that Tyler had discovered would allow the resurrection of many grave secrets locked in stone for more than 65 million years. The presence of rare soft-tissue structures would ensure that this fossil would become a member of a prehistoric elite—dinosaur mummies.

Since I started working at the University of Manchester I have been lucky enough to explore the diverse disciplines that have begun to interface with the science of paleontology. While the days of shovel and spade have not left the field, many 21st-century methods and techniques are becoming available. The excavation process is now treated more like a crime scene, with the macabre secrets of death waiting to be unearthed, analyzed, and interpreted. While we are not greeted with the stench of rotting flesh, cutting-edge science carefully processes the many clues to unlock ancient worlds and the animals that once occupied them.

Every care would be taken to remove, record, and scrutinize the sedimentary blanket that had wrapped Tyler's mummy in a protective cocoon for more than 65 million years. The chemical signatures of environmental change, decay, and mineralization would be painstakingly logged, collected, and shipped back to Manchester for analysis. The samples would be located within a vast three-dimensional digital outcrop map, acquired using the latest laser-scanning technology, helping to place the fossil spatially in both time and changing environment. As the dinosaur fossil was slowly prepared using more traditional paleontological techniques, samples were carefully taken from the delicate skin envelope in the hope that fragile structural biomolecules (proteins and their breakdown products) locked in the mineral-rich skin envelope had survived. X-ray computerized-axial tomography would be used to look into the innermost secrets of our dinosaur, using techniques more commonly applied to the space program. State-of-the-art computer simulations would also be applied to our dinosaur to reveal much of how the animal once walked and ran across the Hell Creek floodplain.

Since the first mummified dinosaur was found 100 years ago, paleontological science now has the potential to transport us back to the final moments of a hadrosaur in the twilight years of the reign of the dinosaurs.

CHAPTER ONE
DEATH OF A DINOSAUR

"One hundred percent of us die,
and the percentage cannot be increased."
—C. S. Lewis

OMINOUS STORM CLOUDS built over the shores of the coastal floodplain behind them. Sweltering moisture-laden air blew through the endless march of dinosaurs. It made the elder members of the herd uneasy, increasing the urgency of the herd's pace. The floodplain had the potential to become treacherous if the rains came early.

The vast herd of hadrosaurs had trodden the same trails for millennia, leaving millions of tracks along the paths of migration. Each step brought them closer to the relative safety of the higher plateau and added security of numbers at the breeding grounds. The *Tyrannosaurus rex* that were continually shadowing the herd had already picked off numerous stragglers over the past few weeks. The herd was vulnerable out in the open, but their sheer numbers worked in their favor. They outnumbered the predators two hundred to one, an intimidating prospect even for the boldest predator. Natural selection was ensuring that only the fittest animals would make it through to the breeding grounds. The hadrosaurs tried hard to remain a tight herd, operating as a cooperative deterrent. However, a severe storm on the plain could cause the lowland rivers to swell. The entire herd could be wiped out by a flood.

While adult hadrosaurs waded through the many watercourses that crisscrossed the vast floodplain with ease, younger animals struggled with the currents and the thrashing bodies of larger, stronger members of the herd. As they crossed each stream, a cautious watch was kept for the crocodiles that lay in wait, still beneath the waters of major crossing points. The crocodiles ate their fill twice a year from the migration of the great herds. While *T. rex* relied upon speed and sheer power, the crocodiles were the ultimate stealth predators, ready to explode from the waters and drag their prey to a watery grave.

The herd reached the main river course, which each year offered a subtly different challenge due to its meandering path through the plain. The herd began to back up on the banks of the wide river. The mist rising over the steady current of the dark, silt-laden waters concealed its immense energy.

A deep, powerful tone resonated from the lead animals. A crossing site had been selected. The first animal, a large dominant male, stepped over the shallow precipice separating the floodplain and freshly deposited sands banked up against the inside meander of the river. The lead male did not sink into the sand, but assertively padded toward the water's edge. He faltered slightly as he reached the edge, knowing his actions would determine the fate of the whole herd. Nostrils flared and bellowing loudly, the 8,000-pound male propelled himself into the water, pushing up a vast bow wave before his broad chest. Soon only his back, neck, and head were visible above the dark waters, as his powerful legs pushed off the soft sediments of the riverbed. His direction of travel was a gentle diagonal traverse of the river, carried by the downstream flow. A steady line of herd members advanced to the edge of the river. The last major hurdle of the season's migration would soon be behind them.

After several hundred of the herd had entered the river course, an explosion of water and teeth erupted as a huge crocodile launched itself from the shallows. A young hadrosaur's head was grasped in the vice-like grip of the crocodile's jaws.

A second explosion followed as another crocodile clamped its jaws around a leg. Screams of alarm echoing along the shore, the herd retreated from the splashing frenzy of jaws, flesh, and bloodstained waters. In a few seconds the young hadrosaur had disappeared from sight. A quiet cloak of death hung over the water. Yet instinct to reach the breeding grounds overcame fear. The crossing resumed.

By midday all but a few stragglers were waiting to take their turn on the now trampled sediments of the river crossing. The crocodiles had feasted frequently for most of the morning, taking dozens of animals, but the numbers of prey left the predators sated and torpid. A young male, not more than ten years old, nervously padded his way to the water's edge. The sweet smell of the water was now tainted with the acrid taste of blood and feces.

As the young male began the crossing, he could already see large numbers of the herd on the opposite bank cloaked in a vast cloud of steam rising from their drying bodies in the midday sun. Plunging in, the animal soon found the river dragging him farther downsteam than was safe. Soon the banks of the river would be too steep for the animal to escape the undertow of the current. Suddenly aware of his plight, the young male began to fight back against the quiet but relentless current. What had started as a tranquil crossing transformed into a struggle for survival. As he pushed desperately with his powerful leg and tail muscles, alarm calls from the opposite banks of the river rose as the herd called to the rapidly disappearing male. As the male was dragged into the vast wasteland of waters where tributaries swelled the river further, the herd disappeared from his view. The exhausted male began to give in to fatigue. The first inhaled water filled the lungs of the hapless animal. In another few minutes the hadrosaur slowly drifted, motionless in the waves.

Another wide meander of the river was dotted with sandbars, a function of the slower moving waters in the inner bend of the watercourse. The fresh carcass of the male hadrosaur settled partially submerged in the waterlogged sands. A small solitary crocodile, a *Borealosuchus*, attempted to gain entry through the tough hide of the animal. The croc, too, would eat his last fill that day. The rains began to fall. The division between river and sky soon merged as the deluge increased. The river slowly rose over the remains of the hadrosaur, rapidly cocooning the animal in a suit of soft, fresh sediment.

Reconstructing a scene that occurred more than 65 million years ago might initially seem far-fetched, but locked within any landscape and its fossils are clues that can help resurrect such places and events. The toolkit that helps unearth these events is similar to that of a crime-scene investigator, albeit the scene of the "crime" is far from fresh.

The processes that affect a body after death are possibly the most difficult secrets to exhume. Within minutes of an organism's demise, the body begins to decompose. The complex relationship between body chemistry and the environment in which it decomposes is one we will explore, as well as the processes of decay. Deciphering the grave secrets of a dinosaur is no easy task, given that millions of years have laid waste to the evidence. Where once a living, breathing organism roamed the land are now only rare, disjointed fossil remains locked in stone. To resurrect a dinosaur from its rocky tomb requires skills that have steadily expanded the science of paleontology.

Eight years ago a prehistoric crime scene was uncovered by a young fossil hunter, Tyler Lyson, that would lead dozens of scientists from many disciplines to start the painstaking process of reconstructing the last steps, burial, and fossilization of a dinosaur in the Hell Creek Formation of North Dakota. The dinosaur, which has been nicknamed "Dakota," is one of a rare type of fossil often called dinosaur "mummies." As used by paleontologists, this loosely applied term has quite a different meaning than that used by archaeologists, yet such fossils provide unique information that has allowed us to fill in many gaps in our knowledge about dinosaurs. This new find represents an exciting step forward in a long legacy of discovery.

Before I focus on this specific find, it is worth a quick review of what constitutes a dinosaur. Many children have come across this famous extinct group in books, on TV, and in the movies. Dinosaurs are a British "invention," although the fossil remains of "antediluvian beasts" were known for centuries, from every continent across the globe. The Chinese have long spoken of "Lung," the dragon, whose bones were scattered across many provinces of China and were often rendered into an apothecary's jar for medicinal use. Such tales were most likely based upon giant fossil bones, sometimes from dinosaurs.

The ancient Greeks and Romans also recorded fossil hunting and the interpretation of their finds. A Roman statesman and general, Quintus Sertorius, in 81 B.C. was reported to have found an 80-foot-long skeleton in North Africa. Pliny the Elder, a first-century Roman scholar famous for his book *Naturalis Historia*, documented the discovery of a 69-foot-long giant revealed by an earthquake, and also described at length the life and habits of the mythical griffins of Mongolia. The area of Mongolia that gave rise to this myth is well-populated by the fossil remains of the lion-sized, beak-faced dinosaur *Protoceratops*. In the third century A.D., the Roman historian Julius Solinus tells of an occasion in the first century B.C. when retreating floodwaters exposed a skeleton nearly 50 feet long in the collapsed sediments of a riverbank.

The Renaissance signified an important step forward in paleontological understanding. Leonardo Da Vinci made many notes on fossil mollusk shells from high mountain passes, speculating as to their origin and preservation. In the year 1565 Conrad Gesner published *On Fossil Objects*, which was one of the earliest attempts to improve on the work of classical scholars. The book was beautifully illustrated with woodcut prints of fossil specimens, a great advance on the often obscure written descriptions.

Some descriptions in Britain from the 17th and 18th centuries carry wonderfully inappropriate names, such as *Scrotum humanum*. Rev. Robert Plot first described the fossil in question as a "human thigh bone" in 1676, but the creature was not named *S. humanum* till 1763 by Robert Brookes. Fortunately, the one with this particular distal end of a thigh bone (femur) now goes by its more recent name, *Megalosaurus*. A number of "saurian" fossils were excavated and named in the early part of the 19th century, such as *Megalosaurus* (1824), *Iguanodon* (1825), and *Hylaeosaurus* (1832), and many more followed. What became increasingly clear to those who studied these bones was that they were an extinct group of animals, unlike anything alive today. What was needed was a suitable means by which to classify these animals.

Richard Owen, a brilliant comparative anatomist, was granted funds by the British Association for the Advancement of Science to explore these fossil remains of "antediluvian beasts." At the 1841 meeting of the Association in Plymouth, Owen presented many of his findings, and these were published in 1842, when he first used the name *dinosauria* to unite this distinct tribe of animals. Dinosaurs had been officially invented. Owen stated that dinosaurs were "a distinct tribe of saurian reptiles," deserving a collective name, dinosauria, literally meaning "terrible lizard." He defined dinosaurs by the possession of five fused sacral vertebrae, the region of the backbone that runs through the hip (pelvic) region. This definition has been somewhat refined and expanded in recent years, now including such characteristics as the possession of a rear-facing shoulder joint and an open hip socket among some specimens.

Dinosaurs were first released on the public, so to speak, in 1854 in the grounds of the Crystal Palace at Sydenham. The "palace," a vast iron-framed glass house, had been moved from its site in Hyde Park, where it had formed the centerpiece of the Great Exhibition of the Works of Industry of All Nations, the first world's fair. A decision was made to populate the grounds of Sydenham Park with life-size models of dinosaurs, marine reptiles, crocodiles, and pterosaurs. The sculptor, Benjamin Waterhouse Hawkins, was commissioned to resurrect this menagerie for the delectation of the public! Dinosaurs were an instant hit, and thousands flocked to see the vast prehistoric reconstructions. Dinosaurs were immortalized by Charles Dickens in the opening passages of *Bleak House*: "Implacable November weather. As much mud in the streets as if the waters had newly retired from the face of the earth, and it would not be wonderful to meet a Megalosaurus, forty feet long or so, waddling like an elephantine lizard up Holborn Hill."

The public obsession with dinosaurs has not waned in more than 150 years. The study of the fossil remains of dinosaurs has expanded into a vast field of research on every continent. From humble, albeit sensational, beginnings in the 19th century to the 21st century with its a vast array of methods and tools, scientists continue to unlock the grave secrets of dinosaurs.

The dinosaurs were subdivided into two distinct groups (orders) by Harry Seeley in 1888. The division was based upon the geometry of their pelvic girdles: the saurischia (lizard-hipped) and ornithischia (bird-hipped). This major division in this group occurred right at the beginning of the Age of Dinosaurs, sometime in the Triassic Period. Some have argued, based on the variation between saurischian and ornithischian dinosaurs, that each has its own distinct ancestor. However, most now agree that these two major dinosaurian orders belong in a single natural group (superorder), the dinosauria. The molecular tools available to biologists to study the evolutionary relationships of modern species are not easily applied to the fossil record, meaning such extinct family trees (phylogenies) are based upon skeletal remains and morphology. Ideally, we would obtain genetic sequences from soft tissues that would allow the reconstruction of molecular phylogenies.

While on the subject of evolutionary classification, it's worth explaining how and why we name all living creatures. Starting with the familiar, paleontologists have a habit of giving a dinosaur a nickname when excavating it. For instance, the *T. rex* Sue and Stan were named after their respective finders; our dinosaur's name was chosen by Tyler, based simply on its geographical provenance. More formally, all types of birds, mammals (including ourselves), fish, amphibians, reptiles, crustaceans, and bacteria have at least two names that follow internationally accepted codes for naming plants and animals. These are the International Code for Zoological Nomenclature (ICZN) and the International Code for Botanical Nomenclature (ICBN). Within these two volumes are the rules of naming a new species. A new name is given only when a plant or animal is discovered for the first time and is shown to be distinct from any known species. The animal—its morphology, fossil, and/or structure—is then formally described and published within a peer-reviewed journal. A new species is then presented to the scientific world, and often to the wider world via the media.

The classification of specific plants and animals into distinct groups or tribes is also worth a quick review. The binomial system applied to naming plants and animals was devised by Carolus Linnaeus in his important work *Systema Naturae*, published in 1735. Linnaeus subdivided a name into first the genus and then species names—for example, we are *Homo sapiens* (always written in italics). The Linnaean system allowed species to be classified within a hierarchical structure, starting with kingdoms. Kingdoms are divided into classes, then orders, which are further divided into genera and eventually into species. The classification is based upon observable characteristics, meaning that many early classifications resulted in very strange family trees. The more specimens and types of characteristics you have, usually the more robust the family tree. However, a new find that possesses a distinct character—say, a theropod dinosaur with feathers—can have an enormous impact that results in major revisions of evolutionary relationships. So, if we were to look at the classification of the species *Edmontosaurus annectens*, it would look like this:

Kingdom Animalia
Phylum Chordata
Class Sauropsida
Subclass Diapsida
Infraclass Archosauromorpha
Superorder Dinosauria
Order Ornithischia
Suborder Cerapoda
Infraorder Ornithopoda
Family Hadrosauridae
Subfamily Hadrosaurinae
Genus *Edmontosaurus*
Species *annectens*

Our own evolutionary path parted with the ancestors of dinosaurs sometime in the Carboniferous Period, when the diapsids (reptiles, and later to include birds) and synapsids (to become mammals) split company. However, the basic tetrapod ("four-feet") skeletal plan is still recognizable, shared by all vertebrates. I love taking my undergraduate class to the Manchester Museum, across the road from the department in which I teach. I show students the five fingers in the hand, the vestigial hips, and the seven neck vertebrae of a sperm whale (*Physeter microcephalus*), and compare those with elephants (*Loxodonta*), antelopes (*Antilope*), and chimpanzees (*Pan troglodytes*). The rough skeletal blueprint is often separated only by form. The evolutionary distance between animals is beautifully displayed by their adaptations, a function of the selective environmental pressures that have dictated their survival.

Occasionally a name is given to an animal that has already been formally described, but has been missed or misidentified by the author. When this occurs, the first name has priority over the second. In some cases this has meant some splendid names have bitten the dust. *Brontosaurus* ("thunder-lizard") was named by Othniel Marsh in 1879, after the remains of *Apatosaurus* ("deceptive-lizard") had already been formally described by Marsh in 1877. The fossil bones were from the same species, so the wonderfully apt name of *Brontosaurus* had to be dropped. The size difference between the two different fossils misled Marsh into thinking he had a new type of dinosaur, when in fact the animals were only at different stages of growth. The misidentification was not recognized until 1903, allowing the term *Brontosaurus* to be well and truly fixed in scientific and popular literature.

The fossil find made by Tyler in the Badlands of North Dakota seemed to be a member of a group of dinosaurs called the hadrosaurs. These are ornithischian dinosaurs; although the unfortunate name refers to the pelvic bones resembling those of modern birds, they are not related. It is the other half of the dinosaur family tree, the saurischians, that are the direct ancestors of modern birds. Saurischian dinosaurs include *T. rex*, *Allosaurus*, *Velociraptor*—in fact, all predatory dinosaurs—and also the vast sauropod dinosaurs, including *Diplodocus* and *Apatosaurus*.

The family Hadrosauridae, sometimes called "duck-billed" dinosaurs, was named in 1869 by Edward Drinker Cope. The fossil remains of this group are found in Upper Cretaceous rocks of Asia, Europe, and North America. The hadrosaurs are further subdivided into the subfamilies Lambeosaurinae and Hadrosaurinae, because only the former had crests on their heads. Hadrosaur fossils were the first skeletal dinosaur remains to be discovered in North America, in 1855, although the tracks of dinosaurs had been known since the early half of the 19th century. Many fossil teeth and partial skeletons of hadrosaurs were discovered over ensuing years.

Several species of hadrosaurs have been found in the Hell Creek Formation: the more common *Edmontosaurus*, the rare and very large *Anatotitan*, and some possible crested lambeosaurine (literally "hollow-crested") hadrosaurs that have not been formally described yet. Tyler was pretty sure that his new dinosaur find was an *Edmontosaurus*. Yet that species breaks down into three subspecies: *Edmontosaurus regalis*, *Edmontosaurus annectens*, and *Edmontosaurus saskatchewanensis*. Only a full excavation would reveal the exact species of dinosaur that Tyler had found.

The name Edmontosaurus was first given to a dinosaur in 1917, based upon some fossil remains found near Edmonton, Canada (the name means "Edmonton lizard"). The first remains of this dinosaur were named *Edmontosaurus regalis* by the Canadian paleontologist Lawrence Morris Lambe in 1917.

Edmontosaurus was a plant-eating (herbivore) dinosaur that is often reconstructed as a biped, walking on two legs, but it almost certainly used its forearms to occasionally walk on all fours as a quadruped. The arms were relatively short compared to the legs, and more lightly built. The restoration of this dinosaur's hands and feet often indicate that it possessed rounded hooves and soft fleshy pads to support the animal's weight. The fossil bone cores that once terminated each toe are broad and flat, indicating that their protective keratinous (nail) sheaths would have been likewise.

The famous dinosaur hunter Barnum Brown examined the hands of a beautifully preserved mummified dinosaur at the American Museum of Natural History in New York and concluded that the fingers of the hand were "partially imbedded in skin." The curator of the museum then, Henry Fairfield Osborn, thought this indicated that *Edmontosaurus* used these specialized hands for paddling around in water and over soft ground. However, Robert Bakker has since argued the webbing was in fact pads from the fingers that had been displaced during the process of mummification. Other workers have also pointed out that the long bones of the hand were tightly bound, as indicated by scars along the bones, not so much an adaptation for paddling as for walking. However, the arms of *Edmontosaurus* are relatively slight, and every time I look at them, they just do not look robust enough for energetic quadrupedal locomotion. This is one problem we will return to later.

The architecture of the backbone suggests that *Edmontosaurus* would have had a low posture, helping it to browse vegetation close to the ground. The low posture would also have dropped its center of mass, improving stability while moving. The vertebrae of the back and tail were braced with ossified tendons (possibly ligaments) that crisscrossed in a trellislike pattern. These structures would have been very flexible in life, not the rigid structures that many authors have attributed to this bracing. If you have ever tried breaking a bone from a freshly butchered animal, you know you have to use all your might to break it, given that it flexes and is incredibly strong. The trellislike structures bracing the backs of hadrosaurs would have had similar properties in life, possibly storing elastic energy to be released from one step to the next.

The skull of *Edmontosaurus* is very distinct. The animal had large nostrils, suggesting it might have had loose skin that functioned as a resonating chamber when it blew through its nose. This might have been used to intimidate other dinosaurs or even attract a mate by visual and vocal displays. While we might find the thought of blowing through our nose to attract a mate unattractive, to hadrosaurs it seems to have been a key adaptation that made them a successful group.

Many species of hadrosaur had complex crests that were extensions of their nasal passages. Some scientists have reconstructed these structures and blown down them to re-create the low-frequency booms these dinosaurs might have hailed each other with. The front of the jaws possesses no teeth, resembling more the beak of a duck than a dinosaur, hence their nickname, the duck-billed dinosaurs. However, inside the mouth at the back of the upper and lower jaws are large tooth "batteries" that contain hundreds of small teeth that were wedged tightly together, interlocking with each other. When the jaws closed, the tooth batteries pushed into each other like multiple chisels, with the lower grinding surface of the teeth sliding over the upper tooth batteries. The skull allowed movement of the lower jaws relative to each other and to the upper jaws, providing a chewing mechanism that was more like the action of a pestle and mortar. This permitted the grinding up of the tough vegetation that would have been a component of the animal's diet. The rounded, beaked front of the jaws would have had a protective covering of keratinous material, making the nose of these dinosaurs even more birdlike. The beak would have helped to cut the tough vegetation before it was curled back into the cheeks and processed by the grinding action of the teeth.

The keratinous beak of hadrosaurs has been observed in stunning fossils like the Senkenberg mummy that we will meet later in this book. Unlike mammals, which mostly have two sets of teeth (milk and adult), the hadrosaurs and all other dinosaurs were able to continually replace their teeth throughout life, with new ones erupting below worn or broken teeth. Hadrosaurs could have up to four or more replacement teeth below the active tooth row, meaning that a single skull might possess more than a thousand teeth. This is a possible reason for teeth being such a common find in the sediments of the late Cretaceous.

It is clear that dinosaurs continue to fascinate one and all. The naming and classification extends our knowledge and understanding of this diverse group of animals with each passing year. The fossil "fuel" on which the fires of paleontology burn is stoked with various grades of fuel, from poor to exquisite preservation. It is the latter that we will explore next.

CHAPTER TWO
OLD FOSSILS AND NEW TAILS

"The most exciting phrase to hear in science, the one that heralds new discoveries, is not 'Eureka!' but 'That's funny. . . .'"
—Isaac Asimov

FOSSILS ARE THE REMAINS or impression of prehistoric organisms preserved in petrified form or as a mold or cast in rock. The term was first used in the mid-16th century, denoting a fossilized fish. The term is derived from the French *fossile* and Latin *fossilis,* meaning "dug up."

The most common form of fossilization produces remains that record only the biomineralized skeletons of prehistoric animals—the "hard parts" of these creatures—since the soft tissues usually decay and disintegrate much more readily. Dinosaur bones and teeth are such biomineralized skeletal features, as are the shells of mollusks, the exoskeletons of insects, and the carapaces of crabs, to name only a few examples. Even biomineralized hard parts can be fragile and subject to decay or mechanical breakdown when they are small in size. The thin and often hollow bones of flying vertebrates such as pterosaurs, birds, and bats are notoriously fragile since they are built so lightly, and they are therefore comparatively scarce in the fossil record. Small mammals likewise have tiny bones that are easily fractured or otherwise broken down and lost to time. Such creatures are most commonly known by finds of their teeth. As with many animals, teeth are the hardest parts of their bodies and therefore the most resistant to postmortem destruction. Many individual animals, especially the early record of mammals and some entire groups, are known *only* through finds of their fossilized teeth.

To extend our understanding of creatures known only through fossil hard parts, we use comparisons with modern life-forms that appear to be structurally similar. This is the science of comparative anatomy, and it has been successfully applied by biologists and paleontologists for centuries. We can also bracket extinct species with living species that are related to ancestral and descendant members of the extinct group. This method of extant phylogenetic bracketing (EPB) was developed by Larry Witmer. In many cases we find additional evidence—such as the scars of muscle attachment on fossil dinosaur bones—that show that the EPB data appear to be appropriate. Combining such evidence and EPB with a liberal application of comparative anatomy, we can build a theoretical understanding of the complete prehistoric animal, as it was with its soft tissues intact.

While evidence supports certain aspects of such extrapolation, in other respects this work is necessarily speculative. About many points concerning soft tissues, we are completely in the dark. The best known dinosaur enigma is their color, which is a particularly ephemeral aspect of an animal due to the nature of biological pigmentation and the tendency for such pigments to be lost in the fossilization process.

Fossilization is a rare phenomenon that occurs to only a tiny fraction of a community's population of any given place and period, and to none at all in many cases. Nonetheless, if we consider the fossilization of skeletal elements as the standard, then the fossilization of soft-tissue structures is much rarer still.

When these unique discoveries are made, this type of fossilization literally "fleshes out" our understanding of the fossil record in many crucial ways. Even a single example of soft-tissue preservation can be of tremendous value in the interpretation of fossil animal types. These discoveries are of such special interest that it is worth reviewing some of the classic examples of this phenomenon. In each case, special circumstances prevented the ordinary loss of soft tissues. The explanations for certain of these situations have been reconstructed with a high degree of confidence, while the reasons for other localities' exceptional preservation remain a mystery. This review will help us to understand just how exciting the Hell Creek find was. Only with this perspective can a full appreciation of these extremely unusual cases be gained.

EARLIEST METAZOANS AND DARWIN'S DILEMMA

The history of complex forms of life begins around the dawn of the Paleozoic Era, some 575 million years ago, when recognizable ancestors of many of today's phyla made their debut. Thirty million years later, a great diversity of new forms of life appeared just before and after the beginning of the Cambrian Period—so many that this burst of evolutionary branching is often referred to as the "Cambrian explosion." This was the great expansion of the metazoans, animal life-forms characterized by their advancement over primitive single cells in several important ways. Metazoans are all multicelled creatures—including yourself—in which cells have differentiated to serve different functions, by forming tissues such as muscles and nerves, and which have an interior digestive cavity or gut. These basic qualities distinguish metazoans from the numberless and mostly unknown single-celled organisms—prokaryotic bacteria and blue-green algae, and later eukaryotic protozoans—which had the seas of the Earth to themselves for billions of years before the Cambrian Period, which began 543 million years ago.

What came before these metazoans that demonstrated such adaptability and evolutionary potential in the Cambrian seas? The earliest multicellular organisms remain invisible to the body fossil record, but trace fossils provide a crucial clue to their existence. Up until 560 million years ago, fossilized prehistoric sea floors show even, regular deposition of fine layers. After that point, however, the layers begin to show disorder and mixing of layers, which is interpreted as *bioturbation*: disturbance by burrowing seafloor animals. Fossil burrows begin to be found, preserved as wormlike casts twisting through the prehistoric sediments, made by burrowing wormlike creatures. Multicellular life had already appeared by the Late Precambrian, although molecular techniques in paleontology push the date for complex life as far back as 900 million years. However, we are fortunate to have extraordinary evidence of several early communities on the ancient sea floor, the Ediacaran biota.

The sudden appearance of the Ediacaran biota in the Precambrian Period heralded the beginning of large complex life. Because these animals possessed only soft tissue, their fossil remains were unknown until a discovery in the Ediacara Hills of southern Australia in 1946 by a mining geologist. The Australian Ediacaran fossils date from 550 million years ago, but since their discovery older localities have been found. The earliest representatives of the Ediacaran biota can be found at Newfoundland's 575-million-year-old Mistaken Point.

The Mistaken Point fossils, called rangeomorphs, represent a group of organisms at or near the base of animal evolution. This is truly where the evolution of animals began on Earth, albeit on the sea floor. The 580-million-year-old Gaskiers Glaciation helps date the Mistaken Point fossils, literally underpinning their stratum. It seems that life got complex immediately following the meltdown of a "snowball earth." Up to this point life had been microbial. What caused the change?

Prior to and including the Gaskiers stratum, sediment chemistry seems to indicate that oxygenless environments prevailed, but in the overlying Drook, Briscal, and Mistaken Point rocks the ratios of reactive to total iron show more oxygenated sediments. This has led Guy M. Narbonne and his colleagues to suggest that complex multicellular life evolved due to a sudden infusion of oxygen into the oceans 580 million years ago. But how the oxygen penetrated to the depths of the oceans and where it came from is still the subject of debate.

A major problem with envisioning these unique Ediacaran fossils is that they are flattened two-dimensional representations of soft-bodied animals that were once three-dimensional. How do you unsquish them? These early animals have been extinct for more than a half-billion years, making their reconstruction very complicated. Such attempts have led to a brisk debate on the evolutionary relationships of these early life-forms. Yet their sudden appearance in the fossil record begs an even bigger question.

Did the Cambrian explosion actually take place? This was one of Darwin's dilemmas. In true Monty Python fashion, Professor Martin Brasier of the University of Oxford would say, "Nobody expects the Cambrian explosion." On a recent trip to Newfoundland I met Brasier and Duncan McIlroy, of Memorial University, to hear about their latest findings on the Ediacaran fossils. Brasier, with a big smile, said much of the activity in this particular area of science was a function of the Mofaotyof principle: "My oldest fossils are older than your oldest fossils." Sometimes science can be driven by the more earthly desires of fame. Nonetheless, the fossils of Mistaken Point have attracted some of the finest minds in paleontology to resolve their mystery.

Brasier has been using state-of-the-art laser scanners that can elucidate the finest details of these squished fossil structures, with many looking like the proverbial Mandelbrot inkblot test. The new laser scanning method has allowed Brasier and his team to unpick the folded structures, reinflating and reconstructing these long-extinct animals in glorious 3-D. These powerful visualization tools seem to be the way of the future. Once we can examine a reconstructed organism, we can also start comparing one fossil to another, and even identify growth stages. Quoting Darcy Thompson, Brasier reminded me with a wry smile, "If you do not understand how an organism grows, you do not understand the organism." The same could be said for many fossil groups that are constructed upon often fragile evidence.

McIlroy, who has been working on the Ediacaran fossils since 1992, was so captivated by the fauna that he eventually moved to Newfoundland in 2005. He started working his way through the literature, local sites, and many Mistaken Point Ediacaran fossils in the collections of Memorial University. McIlroy has made himself at home in the scenic wilds of Newfoundland, with hunting, fishing, and berry collecting a part of everyday life. He is a remarkable chap whom I first met when he was an undergraduate and I a post-grad student at the University of Manchester in 1990. McIlroy has had a meteoric rise in academia because of his direct, well-thought-out, often firm approach to paleontology and geology, based upon his extensive field skills and scrupulous attention to detail when recording data. McIlroy would say, "You have to appraise all the data that are laid before you in the rocks." Specific scientific disciplines have the nasty habit of "cherry-picking" a single component of a problem (the fossils, sediments, geochemistry, etc.) to fit a hypothesis. McIlroy's approach is to use all of the available data. If the data do not support the hypothesis, it's time to rethink the problem and come up with a new solution. In the words of Thomas Kuhn, a paradigm shift is required.

As they work on the Ediacaran fossils of Mistaken Point, McIlroy and Brasier are picking their way through the complex shadows of life that were preserved in the windswept rocks of Newfoundland. They both have realized that these "early" lifeforms are actually quite complex. This supports that the theory that the beginning of complex life was much earlier than the 575-million-year-old fauna they are currently studying. Somewhere in the fossil record lies the answer to this particular question. However, without the Mistaken Point fossils, we would not even know that we have to search deeper in geological time for the moment that life first became complex.

HARD FOSSILS AND BIOMINERALIZATION

The Cambrian explosion was certainly a time of unprecedented evolutionary diversification that laid the groundwork for the body types and basic structural organizations that still dominate the world's biota today. The rapidity of this development and the foundational importance of these ancestral lines of organisms make the Lower Cambrian a period of tremendous paleontological interest. To many evolutionary biologists, this is the single most interesting period in the history of life. For that reason the fossil record of this period is precious to investigators.

As far as the fossil record is concerned, one of the most important features of the Cambrian explosion was that suddenly numerous types of animals appeared whose bodies featured something in the way of hard parts. Up until this point nearly all life had gotten by without becoming any tougher than soft tissue. Stromatolites had formed layered colonies of blue-green algae in reef-like structures and pillow-like lumps in shallow seas since before the end of the Archean Eon, around 2.5 billion years ago. Their fossils, well-known in very ancient rocks, were primitive basic growths that did not possess the biomineralized skeletal elements in individual organisms. In the great expansion of animal types in the Cambrian explosion, organisms commonly formed teeth and gained shells and external skeletons around themselves for protection from the predators that were evolving to ever larger size in the prehistoric seas. The arms race between the hunters and the hunted had begun.

These hard parts—biomineralized skeletons—constitute nearly all of the fossil record, and from such shells and teeth and jaw parts we have discovered many of the creatures that roamed the Cambrian seas. But we realize all too well the limits of the fossil record for this crucial period of evolution. Many of the Cambrian genera we know only by cryptic bits of unassembled hard parts. It is like having the turbine blades from the main combustion engine of the space shuttle and nothing else, and then asking a Victorian ship builder to reconstruct the whole vehicle from which it came.

"Conodonts," for example, was the name given for years to tiny tooth-like fossils that did not seem to belong to any known creature, extinct or extant. The missing "conodont animal" was presumed to be all soft tissue, except for its teethlike elements, and paleontologists hunted in vain for traces of the rest of the mysterious animal. I recall meeting Professor Stewart McKerrow at a Paleontological Association meeting in Cardiff many years ago and talking to him about the conodont animal he had drawn in his wonderful *Ecology of Fossils* book in 1978. He was very proud of his educated guess of what the animal had looked like. In 1983 imprints of eel-like creatures found in Scotland, coupled with conodont fossils in place, proved an educated guess can occasionally be correct. The teeth were indeed serving as dental apparatus of some kind, just as their toothy appearance had suggested. These fossil imprints suggest that some groups of conodont animals were likely members of our own phylum Chordata, animals with backbones (or a notochord precursor of the same). Debate continues about the exact evolutionary relationships of various types of conodont animal, but until the soft tissue fossils were discovered, such questions could not be raised at all. The conodont animals are a classic illustration of the limitations of the fossil record, and the tantalizing amount of information that is ordinarily absent in the rocks.

Charles Darwin insightfully viewed the fossil record as a great library; as you delve deeper in the library, increasingly fewer returns are made, until only occasional sentences and the odd isolated word are left to reconstruct what were once full volumes.

BURGESS SHALE

The Cambrian site that most dramatically demonstrates the limitations of the ordinary fossil record is an extraordinary location called the Burgess Shale. It is sited on the slopes of Mount Wapta and Mount Field in the Canadian Rockies. The Burgess Shale shows just how much we are usually missing, for with this find suddenly paleontology had a complete volume restored to the fossil library.

The site was discovered by Charles Walcott in 1909, a paleontologist who was also secretary of the Smithsonian Institution at the time. In one particular location on a great crumbling face of shale, he came across faint fossil films of ancient seafloor creatures that showed remarkable preservation. Tiny forms, many of them shrimplike in appearance, were preserved as impossibly delicate flattened traces that could be exposed in all their beauty when the shale slabs were split open along their natural bedding planes. Ancestral arthropods, crustaceans, annelid worms, as well as other less readily identifiable fossils, presented not only fine features like delicate antennae in almost photographic detail, but also—more critically—the outlines of these animals' soft tissues, in fine flattened films. Variable reflectivity within these soft tissue films even distinguishes the gut in many of the Burgess Shale animals, improving our ability to identify and reconstruct the probable function of the body structures visible in the fossils.

Walcott excavated this quarry over the next few decades, returning with his family and even exercising the indulgence of naming species and body features of the newly discovered genera after his children. The quarry has continued to be excavated sporadically, yielding thousands of fossils from this unique window in evolutionary time.

The fossil fauna brought to light from this mountainside revealed valuable information about the ancestry and early evolution of the major phyla of animal life. *Santacaris*, for example, proved to be the earliest known chelicerate, the line that would lead to scorpions and spiders. The quarry also produced evidence of entirely unsuspected strange and bizarre creatures. *Hallucigenia*, with its seven pairs of legs and seven pairs of spines, has become somewhat of a mascot of this hitherto-unknown diversity, and the five-eyed *Opabinia*, with its grasping trunk, has so far baffled attempts to identify its evolutionary relationship with better-known lineages. Such are the paleontological riches of the Burgess Shale.

What accounts for the unexampled soft-tissue preservation here, providing this unique view of Cambrian sea life? The Burgess Shale was formed from very fine silts deposited on the sea floor, and fine particulate sediments are ideal for preserving the anatomical detail seen in these fossils. However, the rest of the shale all over the mountainside does not offer such extraordinary preservation, nor do all other fine-sediment shales. The special circumstances that produced the Burgess Shale have now been reconstructed, based on careful analysis of the entire surrounding context of the small site.

A high underwater cliff face stood nearby in the Cambrian, one that can still be seen today, exposed among the darker sediments that eventually covered it. Periodic collapses of fine sediment at the top of this cliff swept bottom-dwelling creatures over the edge. They tumbled down the side in turbidity currents and were buried at the bottom, a hundred feet below. Trapped by the sudden landslides, the creatures were chaotically buried alive and largely undamaged, rather than becoming decayed or half-eaten carcasses when they were uncovered, as would ordinarily be the case. Such a fortuitous circumstance led to the Burgess Shale's remarkably detailed preservation of soft tissues and undamaged, delicate anatomical structures.

Some preservation of altered original biomolecules has occurred in these fossils. Traces of the original material were detected underneath an aluminosilicate film that was produced at the points of body contact by the clay minerals in the silt that smothered the animals. Exactly what prevented ordinary microbial decay in the Burgess Shale specimens is almost certainly a combination of rapid burial, pore-water chemistry, composition of the encasing sediments, and the lack of oxygen.

A comparison of the Burgess Shale fauna with fossils known from the rest of the shale face reveals just how much of the full picture of a living community is lost through ordinary fossilization. In the case of the Burgess Shale animals with hard parts, we have found similar or identical fossils in other localities. But an extraordinary 86 percent of the genera found in the Burgess Shale had no hard parts. A great many of these soft-bodied animals are uniquely found in the Burgess Shale—yet they must have been widely distributed throughout the prehistoric seas. We have here not an unusual fossil assemblage, but only an unusually complete picture of an ancient community. The one component of the community conspicuous by their absence is the juveniles of the species.

The 86 percent figure is a sobering reminder of how limited common fossil evidence is. Even when we are delighted at the abundance of dinosaur bones in the Badland hillsides, the soft-tissue evidence is almost always missing. Yet as the Burgess Shale reminds us, the soft-tissue evidence is a very large part of the picture.

The Burgess Shale stands as a symbol for all that is missing in the "ordinary" fossil record. Truly, any fossil represents an extraordinary confluence of events. Compared to the countless numbers of animals that have lived on our planet, the number fossilized is tiny. When we see hundreds of dinosaur bones at a site, it may mislead us into thinking that the prehistoric world has come to life before us. But it has not—except in the cases of extraordinary soft-tissue preservation, and even then the disjointed sentences of the fossil record can mislead the reader.

LUDLOW: AN EARLY TERRESTRIAL ECOSYSTEM

When working for my master's degree at the University of Manchester, I was luckily employed on the early terrestrial ecosystems project. The project, headed by Paul Selden (an arthropod worker at Manchester) and Dianne Edwards (a paleobotanist at the University of Cardiff), set out to explore key localities in the U.K. that had already provided a glimpse of the world's oldest land animals then. I joined the project after the media hailed their discovery of land arthropods and plant remains in the 414-million-year-old Silurian rocks of the Welsh borderland. I spent many weeks collecting samples of rocks from various sites, but a single site at Ludford Corner in Ludlow (Shropshire) proved to be the most productive.

Each carefully mapped and logged sample was taken back to Manchester and processed in hydrofluoric acid—nasty stuff—to dissolve all rock matrix, leaving only the delicate microscopic organic remains of fossilized plants and arthropods. Each tiny fossil was beautifully preserved, with the plant and animal cuticles looking quite pristine. Some of the most delicate fossils captured gill structures from sea scorpions (eurypterids), which inhabited the marine-laid sediments in which we were hunting for land animals. The land animal and plant fossils had been dragged into the depositional basin by vast storms that had swept across the early land surface, sweeping a snapshot of this flora and fauna into a watery grave.

Before I loosely use the concept of "community" again, it is worth clarifying what this means. Paleoecologists have to view any community carefully, since all of its members are rarely, if ever completely preserved. Conversely, ecologists studying living organisms have the luxury of being able to examine many facets of a living community, not just the disjointed fragments that end up in the fossil record.

The fossils in the Silurian–aged rocks would have clung to the early terrestrial surface, pioneers in the struggle for life on land. Plants had a significant effect on early earth atmosphere and climate, because the proliferation and eventual decomposition of plants caused the first major drawdown of atmospheric carbon dioxide (CO_2), via the chemical weathering of rocks. The calcium and magnesium silicate minerals that were weathered were then subsequently precipitated in oceans as carbonates. That vast store of CO_2 remains to this day as billions of tons of limestone. Plants had started their subtle yet pervasive remodelling of the earth's ecosystems and physical surface.

The backdrop to these momentous terrestrial advances was the death of an ocean. The Iapetus Ocean, which once occupied the Anglo-Welsh basin, gradually closed, eventually leading to the deposition of an extensive bone bed composed of the accumulated remains of millions of early agnathan (jawless) fish (thelodonts). This graveyard was named the Ludlow Bone Bed. It marked the transition from the deepwater marine conditions of the Lower Paleozoic to the continental, freshwater conditions found in Upper Paleozoic rocks in the U.K. However, the change from marine to freshwater in the Anglo-Welsh area did not take place in a uniform manner. The rise and fall of sea levels are clearly recorded in the sedimentary layers.

The Ludlow Bone Bed represents a break in sedimentation, allowing the accumulation of fish remains. An absence of iron enabled prolonged acidic conditions, ideal for the preservation of phosphate-rich bone beds. Due to the fall in sea level, later sediments and fossils indicate a progressively less marine-influenced environment, supported by the presence of salt cracks (syneresis) that signal the encroachment of a delta into the former sea basin. The low pH scenario is backed up by the mode of preservation of the shelly material (i.e., internal casts) and the presence of phosphate rich bone beds.

The microscopic fossil remains of centipedes, scorpions, and early spider-like arachnids (trigonotarbids) were a rare but important find, as were the early land plants that would have created a stable environment for the first terrestrial arthropods to live within. The preservation of plant and arthropod was remarkable. A geochemist at the University of Manchester, Andy Gize, analyzed a sample of the eurypterids using a Gas Chromatography Mass Spectroscopy (GCMS) from the Ludlow site. We were both surprised to find that organic components were still present 414 million years later. While such microscopic time capsules cannot be labeled mummies, they fall into a very select group of fossil sites termed Lagerstatten, the gold standard for fossil sites.

HOLZMADEN

The German paleontological term Lagerstatten denotes deposits that feature unusually extensive fossil preservation of prehistoric life, either in the diversity of genera preserved or in the frequency of preserved traces or impressions of tissues not normally fossilized. The term Lagerstatten therefore applies to some of the localities we have already surveyed, such as Mistaken Point, Burgess Shale, and Ludlow. It also applies to the Posidonia Shale found in and around Holzmaden, a town in southwestern Germany. The Posidonia Shale holds fossils of the Lower Jurassic era, around 185 million years ago. Finds are regularly exposed during shale-quarrying operations that continue to this day. These beds are famous for their preservation of Jurassic marine life such as plesiosaurs, ichthyosaurs, fish, crocodiles, and ammonites.

Ichthyosaurs were a group of marine reptiles roughly contemporaneous with dinosaurs, which had returned to the sea during the Triassic Period and had become fully adapted to a life of high-speed swimming. Their limbs evolved into paddle-like fins, and their overall streamlined shape was beautifully hydrodynamic. The initial Holzmaden finds offered specimens in which skeletons were surrounded by a complete outline of their skin, as if in silhouette. Ichthyosaur skin traces revealed the existence of an unsuspected dorsal fin and a sharklike vertical tail, although, unlike sharks, the endmost vertebrae ran into the lower half of the tail.

The Holzmaden specimens greatly changed our understanding of ichthyosaurs, and paleontologists recognized the degree to which some genera strikingly resembled modern bottlenose dolphins. This is a classic example of convergent evolution, which can produce comparable adaptations to physiological, biological, or mechanical problems encountered in similar environments. In the case of the ichthyosaurs, the soft-tissue evidence was critical to the deciphering of their ecology and appearance, because the dorsal fin and vertical component of the tail fin were entirely composed of soft tissues. The Holzmaden shales have also yielded many ichthyosaurs whose stomachs still contain remains of ingested prey. These include the scales of fish and traces of belemnites (a distant relative of the squid), confirming the ichthyosaurs' prehistoric role as high-speed predators.

The reproduction of ichthyosaurs was formerly thought to resemble that of present-day aquatic reptiles such as sea turtles, which must use their fins to laboriously crawl ashore on beaches in order to lay eggs. The elegant adaptation of the ichthyosaur body form to high-speed swimming, however, appears to have eliminated their ability to maneuver adequately on land, and this would suggest that the animals gave birth to live young like dolphins. The Holzmaden shales helped to settle the question, for among the specimens exhibited today in the Urwelt-Museum at Hauff is a remarkable fossil of a four-meter-long female ichthyosaur caught in the process of giving birth, with four other embryos still contained inside the mother's body. The half-born ichthyosaur stuck in the birth canal must have exhausted its mother during labor; her strength to swim to the surface to inhale air might have failed, causing her to drown with her unborn young. This remarkable fossil captures one of the rarest moments of marine animals, rarely observed in living animals, let alone fossil ones.

AMBER

Fossil amber appears in Greek mythology as the tears of the sisters of Phaeton, who were transformed by their grief into poplar trees as they wept. The Norse called amber the tears of Freya, for after the goddess of love is tricked into infidelity, she weeps tears of gold on land and amber in the sea. In Europe amber has been collected and used for its decorative value since at least Neolithic times, and continues to be popular today. The Roman natural historian Pliny the Elder was the first to document amber, deeming it secreted pine resin that had become hardened. Amber closely resembles the resin exuded by certain trees, commonly appearing as a sort of scab at sites of injury to the bark, or filling hollows in the tree structure. Sometimes mistaken for sap, amber is actually a resin secreted by terrestrial plants' outer cells, which is not the same as the sap fluid, which moves through a plant's vascular system. Various conifer species have been suggested as the sources of prehistoric amber.

Europeans discovered amber in extensive deposits in the countries bordering the Baltic, and these beds continue to rank as some of the finest amber sources in the world. Today the Dominican Republic is also well-known for producing excellent amber fossils, and other significant deposits have been made in Mexico, China, Siberia, Canada, and Lebanon. Any amber-producing locale tends to produce numerous examples of the fossil. The principal amber deposits presently known span more than a hundred million years, the oldest originating 135 million years ago during the dinosaur-rich Cretaceous Period and the most recent deposits having been laid down about 22 million years ago, when dinosaurs had long since disappeared.

I first encountered amber as a child when visiting relatives in Copenhagen in Denmark. I recall the warm glow of amber in windows of shops that had polished vast quantities of the Baltic "gold" and displayed them in a vast array of jewelry. My aunts wore amber jewelry, often chunky, verging on gaudy, and told stories of their childhood when they could pick up the raw material from the beaches around Denmark. I, too, would hunt for amber along the shores of the Baltic each summer, alas with no luck. However, when I was working on the Isle of Wight, I would often find small chips or pitted "tears" of amber in the Wessex Formation of the island, from the Lower Cretaceous. The fragments were too small to polish for potential inprints, but subsequent workers have had more luck at finding "bug" inclusions. Amber is also common in the sediments of the Hell Creek Formation, often appearing in sizable chunks during field excursions and excavations.

Amber, created when externally leaked prehistoric tree resin hardened into a fairly stable form, is of paleontological interest primarily because of its macrofossil inclusions. When fresh and fluid, tree resin is very sticky, and so it tends to trap small animals that alight on it or crawl over it. Further flows of resin from the tree may cover the trapped creature. Subsequent hardening of the aromatic resin as volatile compounds are released into the air can preserve the creature for tens of millions of years, in perfect detail.

In some cases, the organic material decays within the amber and leaves a hollow mold of its original form. In others, some of the more resistant original tissue remains in the amber, especially chitinous arthropod exoskeletal structures, which in turn can preserve within them traces of more delicate tissues. Under high-powered microscopes it is possible to see the tiny hairs on the legs of long-extinct arthropods, that once acted as receptors to their external environment. The detail of preservation is simply exquisite.

Insects are the best known of inclusion species captured in amber and form the most frequently encountered arthropod group. Flies alone account for roughly half of all inclusions. Many other animal groups are also represented in amber finds. Spiders and fungi are among the most commonly found. Occasionally mites, myriapods such as millipedes, and small terrestrial crustaceans such as wood lice are fixed in amber. Other groups include scorpions, mollusks, annelids, nematodes, protozoa, and even vertebrates such as small lizards and frogs, although vertebrates are extremely rare, due to the larger typical body size of these groups. Some birds and putative mammals are known only from amber via feathers and hair.

Plant remains are also frequently preserved in amber, typically in the form of seeds and other small pieces of debris most likely transported to the surface of the resin nodule by wind blowing through the prehistoric forests. Most amber flora fossils are of flowering plants. The preservation of a range of plant species is particularly valuable because we can reconstruct the paleoenvironmental conditions in which the associated species lived, especially when numerous samples are collated from a given locality and a range of flora collected. Most amber-producing paleoenvironments appear to have been warm, wet, dense forests, akin to modern rain forests. Such environments are unlikely to produce conventional fossils because the warm and humid atmosphere promotes rapid and thorough microbial decomposition of animal remains. Amber thus offers evidence of environments otherwise difficult to study in the fossil record.

The preservation of prehistoric creatures in amber has long been recognized as excellent, making amber of interest to nonscientists as well as paleontologists. Amber pieces with insect inclusions are sold as polished specimens to collectors worldwide. Numerous insects have been beautifully preserved, with every tiny feature of their bodies clearly visible. So complete is the preservation in some specimens that detail remains even at the subcellular level. Electron micrography of soft tissues in a Baltic amber midge specimen revealed visible cell nuclei and mitochondria. It is speculated that the sugars in the amber may tend to draw out body moisture from trapped creatures, supporting the extreme degree of preservation that often occurs. In many respects, amber is nature's answer to plastic.

Considering the extraordinary degree of preservation, inclusion fossils became a natural target for the extraction of prehistoric DNA. Successful results have proven that even fragments of DNA are indeed preserved in some amber inclusions. We will return to the delicate subject of prehistoric DNA and biomolecules later.

SOLNHOFEN

During the Jurassic Period, about 155 million years ago, an archipelago of small islands lay roughly at the latitude of today's Florida, in the shallows at the edge of the Tethys Sea—a region that would one day become uplifted into the land we now know as Europe. A fairly dry climate supported a diverse flora of bushes, cycads, and small conifers, but no large trees grew on these islands. Surrounding some of these islands were quiet shallows in which fine-grained limey mud settled to the bottom in an undisturbed environment. The lagoons were protected from wave action by offshore reefs. The arid islands fed little fresh water into these still shallows. The resulting lack of circulation in the lagoon waters created a highly saline and oxygen-poor environment on the muddy floor, so that few scavenging creatures could survive there. Ordinary decay was inhibited. Small animals whose bodies fell or washed into these shallow lagoons might float for a short time before sinking to the bottom, where decomposition was retarded by the hypersaline and anoxic environment. Covered by further fine particles of the carbonate mud, the remains were entombed within a matrix ideal for generating fossils.

Today those fine-grained sediments have become the limestones of the Solnhofen area in Bavaria, in southern Germany. Famous for their smooth structure and fine grain, and their convenient tendency to split cleanly along their pagelike original bedding planes, these limestones have proven useful as tiles and building stones for millennia. They are especially suited to the process of lithography, which was invented in 1798 and developed with the use of slabs of Solnhofen limestone.

The Solnhofen limestones are also famous for the fossils they contain. Just as the fine-grained limestone is ideal for holding fine detail in lithographic printing, so it is also ideal for preserving such extraordinary detail in its fossil inclusions that it has achieved the status of a Lagerstatte.

The dual preservation of plant and animal remains, so rare in the conventional fossil record, has allowed the fairly detailed reconstruction of the paleoenvironment here. Solnhofen fossils are not abundant in these beds, since the Jurassic lagoons supported very little life of their own, and mainly preserved corpses which fell or washed into them from elsewhere. However, when fossils do appear, they are breathtakingly beautiful in their detail.

Quarrying operations regularly bring new fossils to light from the Solnhofen limestones, and a great variety of fossil specimens are now known from this area, including vertebrates, invertebrates, terrestrial plants, and even microorganisms. Some 600 different species have been identified within these deposits. Among the marine animals from the shallow sea waters are a lovely detailed five-armed brittlestar and the "walking" crinoid *Saccocoma*. The image of a soft jellyfish *Rhizostomites* was impressed into the prehistoric mud. Among the insects represented, a *Protolindenia* dragonfly fossil retains its delicate wings with the cell structure still visible. An inch-long beetle that once crawled on the islands ended up in the silt. Pterosaurs probably ate prey just like this dragonfly and beetle. These small flying reptiles are some of the best-represented inhabitants of this ancient ecosystem, with at least 29 different species identified. The Solnhofen beds also claim the honor of the first pterosaur discovered, which came to light in 1784.

Despite these many remarkable fossils, one in particular put Solnhofen on the map: the *Archaeopteryx*, long famous as the "first true bird." A single feather was discovered in 1860, a tantalizing glimpse of the creature it fell from. Beginning the very next year, a series of *Archaeopteryx* fossils began to come to light, at an irregular rate, and ten specimens are currently recognized. From its many dinosaurian skeletal features, its long bony tail, and the fine teeth in its jaws, *Archaeopteryx* might easily have been identified as a small theropod dinosaur, and indeed an amateur mistakenly classified one specimen as the small predatory dinosaur *Compsognathus*. However, the marvelous preservation afforded by the Solnhofen limestones has given us specimens of *Archaeopteryx* surrounded by imprints of their feathers, showing that they looked rather like a modern magpie in their plumage. Wing feathers are asymmetrical like those of modern birds' flight feathers, showing adaptation for aerodynamic use. So detailed are some of the fossils that we can even detect the fine structure of some of these feathers, showing for instance that the fibers of the large flight feathers of *Archaeopteryx* were organized via the barb-and-barbule arrangement that makes modern bird feathers so stable and structurally effective, despite their lightweight and delicate construction.

Several years ago I first visited the Humboldt Museum (Museum für Naturkunde) in Berlin, the home of arguably the most beautiful *Archaeopteryx* fossil in the world. Discovered in 1876 or 1877, it lies on its back with its wings widespread, tail pointing down and head swung over its back. More important, the arms (or, should I say, wings) and tail are surrounded by stunningly clear impressions of feathers. If I had the opportunity to save any single fossil in the world, it would be this one. It is simply stunning. I recall being led into the vaulted paleontology collection store by the curator of fossil vertebrates then, Dave Unwin. We walked to a tall safe in the corner, passing a gorgeous skull of a *Brachiosaurus*. Unwin pulled open the heavy doors of the safe and carefully extracted a large, flat wooden box from the middle shelf. The sunlight from the nearby window glared upon the glass lid to the box, but as he approached cradling his precious cargo, shadows enabled my first glimpse of this incredible fossil. With a grin Unwin announced, "This is possibly the most important fossil in the world." My jaw dropped at the sight of this historic specimen. I spent an hour admiring the intricate tracery of feathers that were delicately impressed in the lithographic limestone.

Ongoing analysis and debate will always persist concerning *Archaeopteryx*—whether it used its wings to glide or flapped its wings to fly like a modern bird. Plus, the exact evolutionary path of this early bird remains uncertain, complicated by the recent discoveries of feathered dinosaurs from the Cretaceous Period.

However, a team led by Angela Milner, at the Natural History Museum in London, recently conducted a CT-scan study of the London *Archaeopteryx* specimen's 20-millimeter-long braincase. They were able to generate 3-D reconstructions that allowed them to investigate the anatomy of the brain in detail. Their study suggests that *Archaeopteryx* was already well equipped for flight, judging by key areas of the brain and sensory systems. "Now that we know *Archaeopteryx* was capable of controlling the complex business of flying, this raises more questions," said Milner. "If flight was this advanced by the time *Archaeopteryx* was around, then were birds actually flying millions of years earlier than we'd previously thought? As yet we have no earlier fossils to help us piece together this fascinating evolutionary story, and of fossils that have captured worldwide attention along with the intense interest of paleontologists. Liaoning's fossils were little known outside East Asia until the 1990s, when dramatically well-preserved dinosaur remains were first brought to public attention. Since then the site has become a paleontological hot spot. Its groundbreaking new discoveries have prompted spirited debate and dramatic revisions of paleontological thought.

The Liaoning deposits, from within the Yixian Formation, yield fossils with dates ranging across a span of 20 million years, in early Cretaceous times beginning 130 million years ago. This period saw the diversification of birds, a growing range of flowering plants, and the development of new kinds of mammals—as well as dinosaurs.

The Liaoning fossils offer an astoundingly rich slice of a prehistoric assemblage. Birds, dinosaurs, pterosaurs, insects, reptiles, amphibians, mammals, and fish all appear in the quarry sites here. Collectively, these fossils allow the analysis of large parts of a prehistoric community. Since 1928 this community has been known as the Jehol biota, after an old name for the region. The remarkable preservation in these rocks makes the Liaoning beds some of the richest in the world.

The sediments of the Yixian Formation tell a turbulent tale of death, disasters, and volcanoes. Active fault lines bounded the basins where the Jehol biota were living. The remarkably preserved fossils were deposited in freshwater lakes, with a regular input of sediments from the nearby volcanoes. The rapid blanketing of the volcanic sediments was a major factor contributing to the exceptional level of preservation and might have contributed to the death of many of the animals. Some horizons within the lake sediments are littered with bodies, indicating mass-death assemblages that repeatedly accumulated after catastrophic eruptions.

The Liaoning fossils sometimes seem like photographic-quality snapshots. Insects, with their small size and delicate nature, dramatize the exceptional preservation at Liaoning. The individual cells in the network of veins across the four large wings of the dragonfly *Rudiaeschina* stand out in stark clarity on one fossil, more clearly perhaps than one would have seen in life when the animal was buzzing about in its hunting. The flowering plants that were undergoing evolutionary expansion at the time required pollination for reproduction, and the Liaoning sediments have preserved some of the insect agents of that vital process. The bee *Florinemestruis* appears in one fossil that preserves even the tiny proboscis the animal once used to feed on the nectar of early Cretaceous flowers.

Vertebrate fossils from Liaoning make rarities commonplace. Amphibians such as frogs and salamanders are generally delicately built animals with fine bones that do not preserve well. They are considered finds of special note in most locales. Liaoning has yielded frogs that are not only complete, but numerous. *Archaeopteryx* was for decades famous as the only fossil type in the world that showed the impressions of feathers. In a fossil of the primitive bird *Protopteryx* from Liaoning, not only can its feathers be seen, but so clearly are their details and texture preserved that three different types can be distinguished: tuftlike feathers covering the animal's head and body, flight feathers on its wings, and midline tail feathers whose appearance betrays their evolutionary origin from reptilian scales.

Not only animals but plants too are preserved at the Liaoning quarries. Most sites favor either animal or plant remains, because the conditions at a site tend to preserve only one or the other. The Liaoning fossils preserve both, and very numerous examples of each. Such a diversity of plant and animal life makes Liaoning more complete in its record of an ancient terrestrial community than perhaps any other fossil locale on Earth. The work of Jason Hilton from the University of Birmingham has helped unlock many of the secrets of the Jehol flora, and many other key paleobotanical sites in China.

Paleontologists are only beginning to delve into the extensive investigations made possible by this sample of unparalleled ecological depth. A Liaoning pterosaur egg, the first ever found and published, in 2004, revealed that these were leathery-shelled. Even the embryo can be evaluated. Its high degree of development indicates that pterosaurs may have been able to fly soon after birth. New fossils continue to be discovered at a prodigious rate, overwhelming the ability of researchers to describe and publish the specimens.

The only regrettable aspect of Liaoning is that the site is poorly guarded, making theft and illegal excavation a constant threat to this extremely valuable record. Diggers and preparators piece together specimens to make them aesthetically pleasing to high-dollar collectors, sometimes destroying vital context in the process. This can even cause serious paleontological confusion, as in the case of *Archaeoraptor*, which in 1999 generated great excitement and media hoopla before it became clear that the single specimen had been made from multiple parts of different animals.

SANTANA

The Araripe Plateau in northeastern Brazil stands above a great limestone formation known as the Santana. It was laid down in the early Cretaceous Period, between 108 and 92 million years ago. Exposures at the edges of the *chapada*, or tableland, expose the Santana Formation (and the Crato, which was originally considered part of the Santana, but is now reclassified by Marto as a separate formation in its own right). For discussion purposes here, the Crato and its contents are included with the Santana.

The first fish fossils from this area were observed in the 1820s, but well over another hundred years would pass before the Santana received intensive study. Late in the 20th century the paleontological study of this area began to catch up with its true potential. A find in 1974 began the recovery of fossil riches from the Santana Formation, which has since earned a reputation as one of the world's rare Lagerstatte.

The sediments of the Santana were laid down at the bottom of a shallow inland sea. Brackish lagoons, somewhat similar to those which produced the Solnhofen limestones, created an environment conducive to undisturbed burial of prehistoric remains. However, unlike Solnhofen, the Santana limestones often preserve their fossils within phosphatized nodules. The formation of these nodules was triggered in each case by the nucleus of animal remains. Since conditions varied over time, these sediments have produced fossils of varying quality.

The Santana preserves a rich sample of the prehistoric ecosystem, with a wide range of life-forms represented in its fossils. Along with dinosaurs, pterosaurs, reptiles, amphibians, invertebrates, and plants, more than two dozen species of fish have been identified. Hybodont sharks, rays, and small coelacanths swam in the shallows. Turtles warmed themselves under the Mesozoic sun. The notosuchid crocodilian *Araripesuchus gomesii* prowled the area. Specimens of this species seen in Brazil have also been discovered in West Africa, striking evidence of the close proximity of the two landmasses at this time, before they were torn asunder by continental drift. Rare frog fossils suggest that Cretaceous freshwater environments existed in the area, implying a wetter environment than that of the Jurassic Solnhofen fauna.

Terrestrial insects are particularly prevalent among the Santana fossils. Damselflies, dragonflies, beetles, grasshoppers, and other genera are evidence of the nearby terrestrial communities bordering the shores of the lagoons and shallows. Numerous individual fossils include even larval forms, allowing for the reconstruction of life sequences among certain genera. Spiders are rare but do appear in the Santana record.

Generally high preservation quality also characterizes this broad range of phyla. Stomach contents are frequently preserved in fish fossils, affording paleontologists an X-ray view of the stomach contents of the fish at the time of their demise. The extremely rare preservation of stomach contents is always of great paleontological interest, since evidence of an animal's diet helps clarify the understanding of its role in a prehistoric ecosystem. Preservation in the Santana is sometimes high-resolution enough to show the individual gills of a mayfly nymph, or soft tissues such as intestinal linings. Other specimens exhibit greater alteration through diagenesis, or change caused by sediment turning into rock, and may be compressed or show detail loss through recrystallization, as in a conifer's pine cone.

Premier among its treasures are the Santana's pterosaurs. The *Anhanguera santanae* is known from a three-dimensionally preserved skull, whence this creature has been given its specific name. Santana pterosaur preservation is often truly remarkable and finely detailed, with some specimens exhibiting traces of muscle and wing fibers. Scholars are debating the outlines of pterosaur wings and whether the skin attached to the body of the pterosaur, which would give them an albatross-like outline, or whether the skin instead ran all the way down to attach to the hind limbs, which would give them a more typically bat-like outline One Santana pterosaur with preserved skin outline gives evidence that at least some genera did indeed have wings with skin attached to their legs.

Clearly, soft tissues preserved in the fossil record can offer a unique insight to the biology, morphology, and subsequent interpretation of extinct plants and animals. Regardless of varying interpretations of fossils from such sites, they do raise expectations of answering questions that were once believed insoluble. The depths of geological time and the breadth of life that has evolved through it provide one of the most enthralling stories the planet has to offer. We often dwell on human history, but that is no more than a single breath of our planet's long life. While the fossil record is by no means complete, the occasional glimpse of "wonderful things" allows insight to this ultimate story of life. Rather than complete books of the "great works of life on Earth," many pages and chapters, if not volumes, are yet to be discovered. The rare insights that Lagerstatten and isolated mummified remains provide the science of paleontology are akin to an exponential leap in knowledge.

The Hell Creek Formation can also be considered one of these unique passages in Earth's history, given its rich bone

CHAPTER THREE
MUMMY

"Shall Life renew these bodies? Of a truth
All death will he annul, all tears assuage?
Or fill these void veins full again with youth
And wash with an immortal water age?"
—Wilfred Owen

THE TERM "MUMMY" has been used to mean many different things. It is broadly recognized as indicating a preserved body, but mummies can be produced in many ways. They can be human or animal. They can be artificially created by human effort, or naturally produced by accidents or particular environmental conditions. A survey of the kinds of mummies will provide a perspective on the kinds of preservation that mummification can entail, and how varying conditions and environments can affect the degree and nature of preservation.

A mummy originally meant a human body from ancient Egypt that had been specially embalmed and treated to resist decay, then wrapped in bandages to help preserve it. The Egyptian practice of mummification was well-known to the classical Mediterranean cultures of Greece and Rome, for it was practiced well into classical times. After the fall of the Roman Empire, knowledge of Egypt was largely replaced by legend. The Arab conquest in the seventh century A.D. further restricted direct European knowledge of the country, and this lent a further exotic atmosphere to anything Egyptian.

When mummies reappeared in European awareness, it was as an export item. The term "mummy" came into our language from the fact that such bodies were, in the Middle Ages, one of the trade products of Arab Egypt. Bitumen, or pitch, is a tarlike petroleum substance that occurs naturally in certain localities around the world, including sites in the Middle East, where it has been used since ancient times for applications such as waterproofing, building mortar, and artworks. Egyptian mummies typically display a darkened skin, and this gave the impression that they had been embalmed with bitumen, even when some had not been so treated. The Arabic word for bitumen, *mumiyyah*, was applied to these bodies, then adopted into medieval Latin as *mumia*, which has become the word we use today.

MUMMY POWDER

By the 12th century, Arab doctors were recommending powder made from mummies for medicinal purposes. While this may seem an unorthodox appropriation of dead bodies from one's own community, which are usually accorded a certain reverence, the Arab conquerors of Egypt did not regard the ancient inhabitants of that land as their own people. This sense of otherness made it easier to consider using these bodies as natural resources for utilitarian applications. Powder made from corpses mummified only by desiccation in sand burials was regarded as a minor medicine, useful for relieving stomach discomfort. However, bitumen was regarded as a more powerful medicine, suitable for applications such as treating wounds, and darkened Egyptian mummies were considered an excellent source. The medicinal value attributed to Egyptian mummies embalmed with bitumen rendered such bodies a valuable commodity, and they were accordingly excavated throughout Egypt for the preparation of mummy powder.

Medieval European doctors agreed with their Arab counterparts and recommended mummy powder widely in Europe as well, creating a substantial market for the product. So large was the demand for mummy powder that, in the 16th century, the supply of easily accessible Egyptian burials had essentially run out.

Suppliers adopted the expedient of fabricating new mummies—obtaining the bodies of criminals, slaves, the indigent, and hospital patients, and preparing the bodies by treating them with bitumen and drying them. An even worse fate awaited 19th-century mummies, with the advent of the railway. Mummies in North Africa were occasionally used to help fire the steam engines. According to Mark Twain, mummies of the richest people (i.e., those coated with lots of bitumen) were preferred as they gave more heat!

The market demand for medicinal mummy powder prompted a hunt for other sources of mummies. It was discovered that the native Guanche inhabitants of Tenerife in the Canary Islands also had a tradition of preserving their dead, especially high-status individuals, which resulted in mummification. Most of these mummies were processed into mummy powder, so this tradition is poorly known today.

THE ARTIFICE OF EGYPTIAN MUMMIES

The oldest mummies in Egypt appear to have occurred naturally. During predynastic times, before 3000 B.C., typical Egyptian burial sites were simple shallow pits dug into the open sands beyond the cultivated fields of the Nile River. The arid environment of the burial dunes inhibited the activity of microorganisms causing decay, as has been observed from intact pre-dynastic burials discovered in archaeological investigations. The sand in direct contact with such human remains helped to draw moisture out and away from the bodies. Progressive desiccation greatly impeded decomposition within these burials, leaving bodies with preserved skin, flesh, and even hair. This is in sharp contrast to the skeletons normally produced from interred human bodies by bacterial decay, which completely decomposes the skin and other soft tissues.

The Egyptian shallow-sand burials might have occasionally been unearthed by sandstorms, or disturbed by predators and carrion feeders such as the jackal. One way or another, predynastic Egyptians observed that the bodies of their dead buried in this way still looked very much like they had when they were alive. The cult belief of an afterlife and the body being the vehicle in which the *Ka* (spirit) made its journey to the afterlife grew, as did the importance of making sure that the journey was successful.

From this natural phenomenon grew the custom of deliberately preserving human bodies as well as possible, along with elaborate religious rituals and beliefs that attended this process. The change to deliberate preservation was probably the result of placing bodies in tombs, since that separated the tissues from the very environment that once preserved them. This necessitated the preservation of bodies through the skills of embalmers. From the early mastabas (simple tombs) to the complex, multichambered tombs of the Valley of the Kings, as the skill of the architects and builders grew, so too did the art of embalming.

The deliberate preparation of human bodies to resist postmortem decay became a high art in Egypt. Other cultures such as the Guanche, mentioned earlier, are now known to have produced mummies, but the process was developed more elaborately in ancient Egypt than in any other culture before or since. The classical Greek historian Herodotus traveled to Egypt around 449-445 B.C. and observed the process while it was still common practice. Herodotus's detailed report of the long procedure involved in producing an Egyptian mummy remains our primary written evidence explaining the phenomena we see in Egyptian burials.

At the height of the art, wealthy individuals arranged for their bodies to undergo a lengthy process taking 70 days, which involved the removal and separate preservation of internal organs, as well as extensive treatment of the skin and detailed bandaging of every finger. Bitumen was one of the outer treatments used to preserve the skin and adhere the protective bandages; aromatic perfumes and spices were also applied. Less well-off individuals had to be content with an abbreviated version of the process, with simpler cocoon-like bandaging. The poor had little more than a bath in salt water and a shroud to approximate to the privileges of the rich.

Natron, a salt-like substance, was a primary element of Egyptian traditional embalming. This material, a mixture of sodium carbonate, sodium bicarbonate, and ordinary salt, was collected from dry, saline lake beds in arid environments where it occurs naturally. Employed as a drying agent, natron also helped produce higher pH levels when in contact with moisture, which inhibited bacterial growth; both qualities served well for mummification. Natron was employed both as a powder packed into body cavities and then removed, and in solutions in which the body would be soaked.

Egyptian pharaohs naturally received the best treatment practiced in their particular eras. Many of these bodies have been roughly treated by tomb robbers and give little sense of the high degree of preservation originally achieved, but in some cases the effects of mummification are still impressive despite crude handling. The face of Ramses II, for example, is so effectively preserved that it still seems to suggest the character of this ruler, dubbed "king of kings" by the English poet Shelley.

Fingernails and hair are commonly preserved by Egyptian mummification, which strove above all for the preservation of external appearances. Internal organs were removed, and some (liver, stomach, lungs, and intestine) were saved in four containers called canopic jars. The brain was regarded as of no importance, as the Egyptians thought that it only produced mucus. It was either pulled out, pulped in the skull by hooks inserted through the nose and then drained out, or dissolved by compounds poured into the skull cavity. The heart remained in the body, since the Egyptians believed that people thought with their hearts, which functioned as the core of their soul, or Ka.

Bandage rolls were commonly used to provide internal support in the place of missing organs so that the skin would retain familiar contours.

These Egyptian practices are salutary reminders that, while a mummy may to some degree preserve the appearance of a living body effectively, external appearances do not necessarily indicate internal preservation. In the fossil record, recrystallization and mold/cast preservation may produce excellent copies of an organism's external appearance without preserving the least trace of its internal arrangements or original biomolecules. Fossil skin and other external soft tissues are often no more than imprints. While valuable for certain reconstructive studies, they should not be confused with actual preservation of original skin tissues.

CHINCHORRO MUMMIES

Along the Pacific coast of South America south of the Equator, environmental conditions produce an exceptionally arid climate in many areas, such as the Atacama Desert, which is one of the driest places on Earth. In the area near the border between Peru and Chile, these hot, dry conditions were exploited by the ancient Chinchorro people for the mummification of their dead. As in Egypt, accidental mummies created naturally by the climate appear to have predated deliberate mummification practices and likely inspired them.

Chinchorro mummification appears to have begun around 6000 B.C. This was a full three millennia before deliberate mummification in Egypt was started, making Chinchorro mummies the earliest known intentional mummies in the world. Early Chinchorro mummies were created by a process that involved disassembling the entire body. Skin was stripped away, viscera were removed, and bones were defleshed and heat-dried. The body was then reassembled and packed with bundles of material such as fur, feathers, reeds, or grasses to fill out the contours, recovered with its own skin (patched with sea lion skin where necessary), further covered in ash paste, and finally coated with black manganese paint. Clay often modeled facial features, which might be topped with human-hair wigs.

After 2500 B.C. an alternative approach was employed, which involved less dismemberment. Evisceration and removal of muscles were carried out mostly through incisions in the skin that were later stitched closed, and the body interior appears to have been dried by the introduction of hot coals before being packed with various stuffing materials. A final phase after 2000 B.C. abandoned the elaborate mortuary preparations and preserved the body through simple natural desiccation before it was given a coating of clay and buried. Locks of hair from such mummies that preserve the original hair allow chemical analysis that has revealed details such as evidence of the chewing of coca leaves through traces of cocaine.

BOG BODIES

Other mummy-like results have been created naturally by the accident of human bodies being placed in environments that have a strong effect of retarding decomposition. Among these are peat bogs and frigid environments. One of the most peculiar modes of natural mummification is that of the so-called bog bodies. However, the use of the term "mummy" is applied loosely here, given that few archaeologists would call a bog body a mummy. Yet the soft-tissue preservation in the finest of these corpses can rival or surpass the most elaborate efforts of artificial embalming and mummification. Eyes, brains, and intestines may all be so well preserved in a bog body as to appear fresh. Bog bodies may consist of no more than scraps or skeletons, but it is the ideal specimens, which feature virtually complete preservation of soft tissues, that concern us here.

Bog bodies are found in the peat bogs of Northern Europe, from Ireland and the United Kingdom to the Netherlands, Denmark, and Germany. The first recorded specimen was uncovered in 1791 in the Netherlands, and hundreds more have been discovered since then as peat in the ancient sphagnum bogs has been mined over the years. As diggers work their way through a level of peat, a shovel will hit something that feels different, or perhaps a hand will come tumbling out of the ragged mass of plant material. Further investigation in the surrounding peat can bring a whole body into view.

Most of the corpses are those of ancient Europeans—Celts and Germanic tribesmen who lived in the northern forests while the Romans lived in the lands to the south. The bog bodies typically date from the Iron Age, within the 800 years stretching from roughly 400 B.C. to A.D. 400. They are commonly given names related to the locale where they are discovered.

Despite their long interment, the bog bodies can display quite remarkable detail. Hair can look so lifelike as to seem that of a sleeping person. Fingerprints may be as clear as those of a living man. Even the "hair gel" used to create Clonycavan Man's pompadour hairstyle can be identified as a particular pine resin that originated in southern France.

Small-scale detail such as fingernails and beard stubble frequently has been excellently preserved in these bodies, while the larger-scale details of posture and volume are apt to be highly distorted by the preservation process and extended burial in the peat. Pressure from overlying strata tends to leave a body flattened. Since most of the bodies do not appear to have been arranged in a standard burial pose, the flattening may produce oddly distorted shapes.

Such distortions of the bog bodies cause no difficulty to modern investigators because we all know what a human's posture is supposed to be, and can "translate" the information from the bog bodies accordingly. However, the warped orientations of their limbs serve as a cautionary tale to dinosaur paleontologists, who do not have live dinosaurs to provide postural models. Paleontologists must cautiously interpret dinosaur bodies that may reflect life posture only as seen through the distorting lens of taphonomy and diagenesis, two of the processes of fossilization.

The desiccation of spinal ligaments in a long-necked animal, for example, often pulls its head back in a classic dinosaur pose that has little to do with the animal's posture in life. Valuable clues as to posture may remain in heavily distorted bodies, but they must be interpreted with care.

BOG PRESERVATION PROCESS

The preservation of bog bodies can include not only the fine details of their skin and hair, but their internal organs as well. Oldcroghan Man, a recently discovered Irish bog body, contained residues in his gut that researchers were able to identify as the traces of a last meal of buttermilk and grain products. What accounts for this type of preservation is the chemistry of the bog environments. These wetlands feature soggy, dense growth of plants such as sphagnum moss, which become peat when compacted by overlying layers over time. Bog waters are suffused with tannins (organic acids) and even aldehydes, which act to kill microorganisms, inhibit bacterial decomposition, and promote the preservation of soft tissues. So rich are bog waters in dissolved tannins from the plants within the bog that the waters are often stained brown by the tannin. This same brown appears in the skins of bog bodies. The ambient chemicals in a bog can act much like the tannin derived from bark that leather workers use to tan hides. The bog bodies are essentially tanned into leather by their immersion, which accounts for their leathery appearance as modern specimens.

The biochemistry of bog preservation becomes of particular interest as paleontologists seek to reconstruct the biochemical situations responsible for the state of preservation seen in dinosaur mummies: A look at modern processes may offer critical insight into prehistoric conditions. However, on closer analysis the precise nature of bog preservation has proved to be complex. Several different models have been proposed to explain the anti-microbial and acidic environment in bogs. Whether the effect arises from the ambient acids, or from the depletion of essential metal cations and amino-nitrogen by sphagnam in the moss, from fungicidal sphagnol released by the moss, or from other factors has not been conclusively demonstrated yet. Indeed, the variations in preservation of bog bodies, which are not all alike, may reflect varying conditions in the bogs, and different factors may come into play under different circumstances.

Most of the bog body finds cannot be considered graves, since most are sacrifices and/or executions with bodies thrown in the bog. The customary treatment of bodies during this period among the Germanic and Celtic peoples was cremation, so the bodies in the bogs do not represent members of a cemetery. Rather, the bodies are typically isolated finds. This cryptic scenario has invited much speculation in the history of exploring the nature of these distinctive human remains.

FORENSIC ANALYSIS

Examination of the bodies has often shown various pathologies—injuries such as broken bones or evidence of disease. Earlier interpreters postulated that people preserved as bog bodies were most often killed violently, in many cases at the very site, and then immediately thrown into the bog or a pit therein. Scholars accordingly proposed that the bodies must represent criminals, murderers, or other social outcasts. The ancient Roman historian Tacitus wrote *Germania*, a contemporary account of the German tribes in the first century A.D., in which he described the Germans' capital punishment of criminals and outcasts by staking them in the bogs.

Recent scholarship and forensic investigation has suggested alternative explanations for injuries in bog bodies. Postmortem damage inflicted during removal from the peat bogs is more likely to account for many recorded "injuries" such as broken limbs. In some cases the modern damage is obvious, such as that to Yde Girl. She was recovered by peat dredgers in Holland in 1897, and the body has since been identified as that of a 16-year-old girl who died somewhere between 200 B.C. and A.D. 200.

The body bears severe mutilation—from the iron tools used by the dredgers. Even the cracked skull of Grauballe Man, says a Copenhagen team led by Neils Lynnerup, may have occurred simply due to the pressure of overlying strata or the footstep of a clumsy gawker who trod where the fragile head was soon to be uncovered. Still, Grauballe Man did have his throat cut open from ear to ear before he was deposited in the bog site. Oldcroghan Man had his nipples cut and twisted hazel thrust through holes cut in his upper arms. A body found in Bourtanger Moor in the Netherlands in 1904 had its intestines partially emitted through a stab wound. Elling Woman, found in 1938, was found with the leather thong used to hang her still in place, while Yde Girl still bears the woolen band apparently used to strangle her. Such instances attest that the traditional scenario of violent deaths often has some basis in fact.

SPECIMEN DAMAGE DURING REMOVAL

Lynnerup's work does make one think about specimen-removal damage. Shovel and spade excavation by hand was the usual way in which bog bodies were formerly discovered, and as a result the few bodies that survive from these older discoveries are often in excellent states of completeness. Grauballe Man, discovered in 1952 in Nebelgård Mose on Jutland in Denmark, is a prime example of such superb completeness, as is the head of Tollund Man, discovered in 1950 in Bjældskovdal, also in Denmark. By contrast, mechanical extraction and mining of peat is the norm today, and so bodies that are spotted during the course of such operations are usually fragmentary. In 2003, Clonycavan Man, from County Meath in Ireland, was discovered in the sieve of a peat-processing plant, but only his torso and head were recovered. His lower arms, hips, and legs went through the peat processing, unfortunately. Due to industrialization and population growth, the remaining old peat bogs are quickly being drained for development and the peat processed by machines and factories, so the future discovery of excellent specimens is likely to become rare indeed. Analysis has indicated that almost all of the bogs that formerly existed in Europe are nearly exhausted, and that we must therefore presume to have run through nearly all the bog bodies we are going to find.

However, in 1984 at Lindow Moss near Manchester, a body was found. He was in fact the second "body" to be found, as the head of another individual, also male, had been discovered earlier in 1983. He was affectionately named "Pete Marsh," and radiocarbon dating suggested that he had died in the first or early second century A.D. At first he looked like a murder victim. X-ray analysis showed that Pete had been hit on the head with a blunt instrument, garroted (breaking two of his neck vertebrae), and then his throat had been cut—a thorough job worthy of the Sopranos. The thin rope that was used to strangle him was still wrapped tightly around his neck. His manicured nails have been interpreted by many to suggest that Pete was of high status, but this is difficult to prove with little else associated with the body. However, his last meal, albeit paltry in quantity, was still inside his stomach after 2,000 years.

Again, the nature of retrieval has its counterpart concerns in paleontology, in that it illustrates how heavily dependent scientific investigators are on the work of the collectors—the excavators, extractors, and preparators—in order to collect their analytical data. A great deal of data can be lost in improper excavation, either by looters or by other diggers who are not skilled or cautious enough, or are in too much of a hurry. With dinosaur fossils, only a well-informed person can appreciate just how much may be lost in a given specimen.

X-RAY VERSUS CT SCANNING IN SOFT-TISSUE ANALYSIS

The same depositional environmental chemistry that so effectively preserves the skin and flesh of bog bodies tends, in contrast, to dissolve the hard tissues of the skeleton. The wet environment liberates the minerals and transforms bone into soft and rubbery collagen. This critical molecule gives all bones their flexibility and much of their composite strength. Twentieth-century analysis of bog bodies employed x-rays in the hopes of achieving non-destructive investigation. Grauballe Man is extremely well preserved and has accordingly been extensively studied, but the demineralized bones of this specimen registered only faintly in the radiographs. CT scans have proven more effective at revealing the inner structures and tissues of such specimens.

REFRIGERATION

In 1991 German hikers in the Ötztal Alps, near the border between Austria and Italy, came across a body weathering out of glacial ice in a mountain pass high above the snowline. The body, only having been partially exposed, looked like a skeleton covered with a wrapping of skin with some flesh beneath. Presumed to be a modern corpse, the body was roughly removed with an ice axe and jackhammer by the mountain rescuers. Only later did investigators realize it was a body of great antiquity. Subsequent analyses dated "Ötzi" to 3300 B.C., from Europe's Copper Age. This was a remarkable find, from a rare slice of archaeological time. In all fairness to the Austrians who recovered Ötzi, modern bodies are much more common than ancient ones and can look much the same.

Ötzi died while crossing mountains in the Alps and was covered by subsequent snows. Refrigeration preserved the body excellently, and indeed it appears that some of the damage it suffered was incurred during and after its removal from the ice. The ice axe used in removal penetrated the body's hip, and the body was not given particular care at first because its true value was not suspected. An amateur video of the "excavation" makes brutal viewing, showing the body of Ötzi being forcibly dragged from his icy tomb with very little ceremony.

Cases such as this are powerful reminders of the importance of vigilance and extreme caution when dealing with specimens of exceptional preservation. Removal is very likely to cause damage unless steps are taken to forestall every foreseeable injury.

The eagerness to expose a well-preserved specimen has often cost science dearly in exposing the specimen too soon to weather, contamination, or unfavorable conditions. The Siberian Ice Maiden was allowed to melt partially before the body could be placed under controlled refrigeration, and Ötzi endured various indignities and exposures before proper protective measures were taken. The excavation of this body was not archaeologically controlled, so context and some of the surrounding artifacts were lost. Later excavations of the site recovered much of Ötzi's belongings, including a copper axe, flint knife, a quiver of arrows, and a long bow, among many other items of clothing.

The preserved body of Ötzi has nonetheless afforded an unprecedented look at an individual from the Chalcolithic Period of Europe. So extensive is his preservation that a great battery of analyses and medical investigations have been carried out, revealing great detail about this individual's life and physical condition. Pollen analysis has been particularly productive. Fresh pollen recovered from one of the two meals still remaining in his digestive system shows that it was consumed (probably accidentally as a wind-blown addition to the meal) in the spring. This data combined with evidence of einkorn wheat bran and blackthorn kernels support the inference of food harvesting and storage behavior, since einkorn wheat is harvested in late summer and blackthorn in the autumn. Ötzi's intestines show that he carried whipworm; parasites are one of the most common finds in mummies of all kinds. The positioning of his last meal in his digestive tract—in the transverse colon—allowed researchers to determine that it was consumed about eight hours prior to the man's death, and the pollen inclusions show that he visited a valley to the south of the mountain location where he died. The cause of his death was remarkably overlooked in the first radiological studies, but the head of an arrow was found lodged in his shoulder by later workers utilizing CT scanning techniques. The wound likely was a contributory cause to Ötzi's fate. The combination of blood loss, altitude, cold, and bad luck led to Ötzi's demise.

The numerous pathological analyses, as in the case of many mummies, helped to build a picture of the injuries, illnesses, and nutritional development experienced by this Chalcolithic Alpinist, but these approaches can be applied even to mere skeletal remains in many cases. We focus here on some of the unusual analyses made possible by the special preservation condition of mummification.

THE GREENLAND MUMMIES

Mummies can provide surprising insights when their special preservation is creatively exploited. Eight Inuit (Eskimo) bodies were discovered in two grave pits at Qilakitsoq, Greenland, in 1972. The two brothers who discovered the graves while hunting alerted the authorities, who did not excavate the site until 1977, when it was found to date from about 1475. Protected from the weather by their burial, the bodies had been largely preserved by the area's climate, which maintained them under low temperatures and low humidity—effectively freeze-drying the bodies and all their many trappings, for they were buried fully clothed.

Isotopic analysis on human and animal collagen can parse out three major groups of dietary sources based on the ratios of naturally occurring carbon isotopes. Two major groups of land plants differ in their use of either three- or four-carbon molecules in carbon-dioxide fixation, while marine plants fix their carbon in another distinctive fashion. Each of the three groups leaves a distinctive chemical signature in human and animal collagen since their food chains ultimately trace back to plants. Through isotopic analysis via mass spectroscopy, collagen from a four-year-old boy among the Greenland mummies revealed that his diet was based 75 percent on marine sources (such as pinnipeds, cetaceans, and fish) and 25 percent on terrestrial sources (such as ungulates, small mammals, and land plants).

One of the Greenland mummy bodies contained fecal material which, when analyzed, was shown to contain the pollen of mountain sorrel. The pollen of this plant is only produced in the months of July and August, which gives the season in which the fecal material was produced, and therefore the season in which the individual's death occurred.

THE SIBERIAN ICE MAIDEN

Cold also helped preserve an ancient Pazyryk woman, dating to the 400s B.C., whose elaborate undisturbed burial mound was discovered in Siberia in 1993. The body had been embalmed and packed with peat and bark, which would have provided some preservation through its tannic acid content, but the primary factor in her preservation was flooding, which appears to have occurred soon after the interment. The waters then froze, locking the tomb in permafrost for 2,400 years. This preservation has left the Ice Maiden's hair still blonde, and she bears elaborate and fantastical vivid blue animal tattoos on her still pale skin. Pazyryk mummies of this type have been found in other tombs in the region, which were protected from the sun's heat by rock piles and thus kept in cold storage after their initial burial. The mummies exhibit careful embalming that involved evisceration and defleshing, followed by restitching of the skin using horsehair thread.

Horses preserved in the tomb with the Ice Maiden appear to have been sacrificed as part of the burial. Dental examination shows that all six horses were old, rather than strong and vital animals that would be a much greater loss to the community. The horses had not been embalmed, but had been well preserved by the natural freezing, to the degree that their stomachs could be sampled (by researchers sufficiently determined to brave the smell, which was reported still to be very strong after two and a half millennia). The presence of a horsefly larva in one stomach conclusively dated the horse's death to the later two weeks in June, the only time of year when this species of fly is at its larval stage. Since the horses appear to have been killed specifically for the burial, this provides a probable time of year for the event, a striking degree of precision for an event so far in the past.

mummies. The *Mammuthus primigenius* that wandered the arid steppe grasslands of Siberia during the last ice age did so recently enough that mammoth ivory has long been found widely in the region. Prehistoric mammoth ivory has been an export commodity from Siberia for more than two thousand years, and mammoth tusks are still actively harvested today. The ivory tusks are comparable to the eggshells found in Madagascar of extinct *Aepyornis*, "elephant birds," in that the tusks and shells are remains produced by animals living recently enough that these biomineralized hard tissues have not deteriorated. Most mammoths died out around 10,000 years ago. However, a small population that was literally small, of pygmy mammoths, survived in isolation on Wrangel Island off the coast of Siberia until about 4,000 years ago, when they finally became extinct. It is an interesting thought that when the Great Pyramid of Giza was being built, mammoths were still alive and kicking in the far north.

While mammoth ivory is far more common than most people realize, preserved mammoth bodies are far more rare. Fewer than a hundred mammoth specimens have been discovered in Siberia, if one counts skeletal remains as well. From this total two dozen complete skeletons have been reconstructed. Of the approximately three dozen mammoth carcasses ever found, most were only partial. Only about four were substantially complete.

The accounts of mammoth bodies encountered in the remote tundras of the far north have been embroidered and exaggerated. Popular impressions suggest completely preserved specimens that have only recently collapsed. Some stories recount that the mammoths' meat was still edible when found, and that discoverers dined on the Ice Age flesh. Such accounts fall under the category of tall tale. In every instance the flesh exposed begins to decay immediately, and where it exists in any quantity it carries a rank odor hardly conducive to dining, except for dogs and scavengers. A scientific team in 1901 seriously considered tasting the inner meat from a mammoth mummy, which looked just like fresh meat, but none of the team could bring themselves to risk it.

THE ADAMS MAMMOTH

The first mammoth brought back from the Siberian wilderness to civilization proved hardly less surprising than a stuffed Abominable Snowman would today, confirming what had until then been merely outlandish legends and folk tales of great furry beasts that lived underground in the treacherous lands of the tundra. On the coast of the icy Laptev Sea in remote north-central Siberia, a local hunter named Shumakhov noted an odd large lump in the frozen ground in 1799. He watched it emerge from the shoreline as the natural thawing progressed, little by little each year, until in 1803 the entire carcass of a woolly mammoth collapsed out of the permafrost onto a sandbank. Shumakhov cut off the tusks and sold them for the ivory content. Pleased with his find, he related the story of their origin to the merchant who bought the tusks.

In 1806 the Scottish botanist Michael Adams heard accounts of the alleged shaggy hulk from which a certain pair of excellent tusks had come. Convinced by further inquiry that the wild story had some basis in truth, Adams made the long, arduous journey to the Laptev coast site to see for himself. The locals had stripped the carcass by this time, carving off the meat for their dogs. Carnivores and scavengers had eviscerated and defleshed the rest of the body. Nonetheless, the skeleton and most of the woolly hide survived. Adams gathered up all the shredded and torn fur he could find from the scavengers' leftovers, and brought it back with the rest of the remains. The St. Petersburg Institute of Zoology museum received the Adams find, creating the first mount of a woolly mammoth. Adams's find also succeeded in transforming a myth into the basis of a new understanding of the history of life.

Mammoths appear to have been preserved most commonly in cases when the animal became mired or sank in a treacherous wet environment such as a bog, the muddy shores of a lake or river, or a frozen lake whose ice the mammoth broke and fell through. In each case, the body became concealed and locked in the permafrost. Riverbanks contain a high proportion of mammoth remains, which may have been preserved during times of floods. This combination of entombment and waterlogged refrigeration forms the standard scenario for mammoth preservation, rather than the open-air freeze-drying, which preserved bodies such as the Greenland Inuit mummies.

Mammoth mummies are discovered when some factor causes part of the body to become exposed. Changing deposition and erosion along rivers and streambeds have exposed bodies that were previously buried, just as the Adams mammoth was uncovered. A century later, the great Berezovka River mammoth mummy was exposed by the collapse of cliffside sediments near the Arctic Circle. This mummy was reached by a scientific team before much damage occurred beyond the flesh and skin of the head. Removed little by little, the mammoth was carefully butchered in place for delivery to St. Petersburg in 1901, where it is mounted today in its death pose just as it was found, with a reconstructed head and trunk. Unusual thaws and warm weather can lead to fresh exposures of bodies that have lain concealed under ice or snow, as in the case of 18,000-year-old mammoth remains recovered in 2002 and later displayed at the 2005 World Exposition in Aichi, Japan. Exceptionally low water levels can reveal sediments ordinarily immersed, and expose mammoth remains as well.

When mammoth bodies are discovered as natural exposures, they begin rapidly to decay and putrefy. Frozen in an arrested state rather than freeze-dried into a permanent state of preservation, their flesh retains all the potential of a fresh kill for supporting standard bacterial decomposition, as well as attracting macrobiotic scavengers such as the bears, wolves, and foxes that fed on the flesh of the Adams mammoth. When it begins to thaw, a frozen mammoth mummy quickly becomes just a moist carcass.

Modern mammoth hunters therefore look for small exposures—a few vertebrae poking out of the ground, perhaps—in hopes of being able to uncover a preserved mammoth before natural exposure occurs and ruins the rest of the carcass. Ground-penetrating radar is employed by some searchers to determine whether enough of the mammoth remains underground to warrant the difficult work of excavation in this environment, where time is a critical resource. Mammoth hunters must carry out their activities during the short Siberian summer before the cold once again locks everything down.

The preserved soft tissues of mammoth mummies offer anatomical information unobtainable from the skeletons alone. The fur shows the color mammoths were (in contrast to the stone fossils that never tell us what color dinosaurs were). They had grayish-brown skin, with a camel-brown undercoat of fur covered over with reddish-brown outer hairs. The mammoths' trunk structures are revealed by the mummies, ending in a curious, almost hand-like combination of elements, with a straight lower lip edge like a paint scraper opposed by a single central finger-like projection above. Mammoths have only four toes rather than the five of modern elephants. They even feature a quite unsuspected adaptation to the cold of the Ice Age: a flap of skin protecting the anal opening. Tusks discovered in place show that these spiraled inward, toward each other, rather than outward, as had been imagined. Their diet has been confirmed from the leaves, grass, and buttercups discovered still in the mouth, as well as in the stomach.

DIMA THE BABY MAMMOTH

On June 23, 1977, the Siberian strip miner Anatoly Logachev was bulldozing a patch of freshly thawed ground when his machine encountered a tough, dark hairy mass in the earth. He dismounted to investigate, and his comrades helped him uncover the anomaly by hosing the area with warm water. Logachev had discovered the complete mummy of a baby woolly mammoth. To interrupt mining operations would cost him part of his livelihood, but for the sake of this remarkable find, Logachev did stop, and undertook to preserve the carcass, which acquired the name "Dima" from a nearby stream.

Because the body had been entirely covered, it was a very unusual discovery. The furry little beast was complete and in an excellent state of preservation. Though Logachev did not know it, he had discovered the finest mammoth mummy ever found. Blondish fur covered Dima's body, with the darker and redder adult hairs beginning to grow out. Analysis would later place the baby's age at six to eight months when it died. Dima seemed as if he could have been recently alive. Traces of his mother's milk remained in his stomach. He had stumbled into a mud pit and eventually succumbed there, sinking below the surface.

Dima's mummy had survived 40,000 years underground, but as soon as it was exposed to air, the clock started ticking. The carcass began thawing in the Siberian summer warmth at the mine site. Decay set in, and flies came. Logachev and his co-workers took the best measures they could: They built a tent to shield Dima from the sun, and they covered the body with ice to keep it chilled.

After three days of exposure, Dima was taken into protective custody by Soviet authorities. The mummy was brought to Leningrad for treatment by preparators at the renowned Institute of Zoology, under the best mammoth experts in the Soviet Union. Regrettably, the Institute preparators soaked Dima in benzene and then embalmed the body with paraffin, a one-two punch that removed almost every trace of hair and blackened the skin from its natural light brown to the color of pitch. Again the lesson is illustrated that conservation and preparation are critical aspects of special-preservation finds. As is so often the case with exceptional specimens, this one would have required exceptional care to preserve its phenomenally complete suite of data. We can only hope that future finds will meet with more fortunate treatment.

THE WOOLLY RHINOCEROS

Mammoths are not the only Ice Age species to be naturally preserved in forms that can be considered mummies. An Ice Age contemporary of the woolly mammoth was the woolly rhinoceros

(*Coelodonta* sp.), which appears in beautiful Paleolithic cave paintings in Europe. The animal is known from skeletal remains, but these all pale beside one quite extraordinary find from Starunia in the western Ukraine. An adult female woolly rhino was discovered here in 1929, in a very unusual preservation environment: A so-called tar pit within an ozokerite mine was found to hold the nearly complete body of a woolly rhino. Although the beast had lost its fur (and its hooves), the soft tissues were protected from decay by the microbe-hostile environment created by the salt and oil in the surrounding sediments. Ozokerite is a petroleum-related mineral wax. This unique specimen can be seen in the Natural History Museum at Kraków in Poland. Only in 2004 were new expeditions to the remarkable Starunia site considered, and it may yet yield further finds of interest, perhaps even new mummies.

Still other examples of natural animal mummies could be cited, such as the occasional desiccated seal that turns up in dry Antarctic valleys, but these instances give a sense of how rare natural animal mummies are, and how extraordinary conditions must be to allow them. This survey has also suggested some of the special scientific value of soft-tissue mummies. It is only with an understanding of the phenomenal rarity of natural animal mummies that one can fully appreciate the significance of soft-tissue fossils and animal mummies from the greater depths of prehistory.

CHAPTER FOUR
DINOSAUR MUMMIES

"Generations to come will scarce believe that such a one as this walked the earth in flesh and blood."
—Albert Einstein

IN AN IDEAL WORLD DINOSAUR FOSSILS would include complete soft-tissue preservation, including organs, sinew, and bone. Unfortunately, such a scenario is impossible due to the taphonomic (decay) processes, which greatly affect an animal's carcass in the wilderness before it can start the long road to becoming a fossil. However, the rarest dinosaur fossils have approached, albeit distantly, this theoretical ideal of complete preservation: the so-called dinosaur mummies.

Dinosaur skin impressions occur in various contexts. Just as dinosaur bone fossils may turn up in isolation, scattered from their original position with the rest of the body due to any number of postmortem factors, from scavenging to weathering, fossil dinosaur skin impressions may also be found as isolated patches rather than associated with a dinosaur skeleton. In other cases, skin impressions are found only in bits and pieces on a skeleton—for some reason, patches of skin on hadrosaur tails are the most common. A helpful starting point for the definition of a dinosaur mummy was proposed by paleontologist Kraig Derstler of the University of New Orleans: "When dinosaur skin impressions are extensive enough to wrap around a substantial amount of the articulated skeleton, the specimen warrants the name 'dinosaur mummy.'" I would argue that this definition is not complete enough, given the term implies that soft tissues are preserved also. A possible addition to Derstler's definition would be: The fossil must include original biomolecules or their decay products either within the bone and/or soft-tissue structures preserved. The only problem here is that such a definition might severely limit the number of "true" dinosaur mummy fossils to possibly one, the subject of this book. So I will follow Derstler's definition and examine some of the fossils that have met the existing benchmark.

Only a handful of paleontologists and fossil hunters have ever known the experience of recovering a dinosaur mummy. Their stories illustrate the changing nature of paleontological science as new concerns and new analyses are brought to bear on these most special of dinosaur fossils. But the stories of dinosaur mummies also illustrate the fact that field paleontology remains in many ways similar to what it was more than a hundred years ago. Large, delicate objects must still be separated from their rock matrix and pulled out of the ground without damage, and then transported from the field to a laboratory or museum.

THE STERNBERGS

In the history of American field paleontology, one famous "dynasty" of fossil hunters takes a prominent place, and in the history of dinosaur mummies, that same dynasty stands unquestionably matchless. This family of remarkably productive fossil hunters consisted of a man and his three sons. To fully appreciate these individuals and their discoveries, we must journey back to the very different era to which they belonged.

The later 1800s were the days of the legendary American "Old West," when the plains, mountains, and deserts of the Louisiana Purchase territory and beyond to California were populated by pioneers, cowboys, the U.S. Cavalry, Native Americans, and vast herds of North American bison (*Bison bison*: often incorrectly called buffalo). Although hard to believe, during this era of gunfighters and frontier expansion, fossil hunters roamed the prairies and badlands just like the cowboys. This was in fact one of the great eras of field paleontology. The first half of the 19th century may have belonged to Britain in the so-called "heroic age of geology"; the same can be said for North America in the second half. The fossil "bone rush" of North America was in full swing.

The prospectors of this time faced a vast open landscape filled with the promise of finds that would begin to flesh out the great unknowns in our understanding of the past. From today's vantage point it takes real effort to imagine how many discoveries lay before these early field workers. Entire major groups of prehistoric animals and dinosaurs were being brought to light for the first time as these explorers investigated truly Wild West areas never before subjected to paleontological study.

Two great paleontologists dominated this era in the United States, and they are as famous for their feud as for their finds: Othniel Charles Marsh and Edward Drinker Cope. Very different as individuals, the two were larger-than-life characters who both led expeditions into the field themselves and dispatched teams working on their behalf. The adventures and conflicts between these two "dinosaurologist" legends could fill an entire book (and indeed have filled several). Their men in the field spied upon one another, and rival teams reportedly came to blows over fossils at times. Based back East, Cope and Marsh sent expeditions out into the Wild West in search of new and more spectacular finds, each hoping to best the other with discoveries of larger size or by finding greater numbers of new species. The competition invested their efforts with great energy. With extensive financial backing available to both men, they drove scouts and prospectors and excavators who returned tremendous benefits to the science of paleontology and the study of dinosaurs in particular.

In the summer of 1876, the Battle of Little Bighorn saw the Sioux famously victorious over Gen. George Custer and his

Seventh Cavalry. During that very same summer, and in the very same region—southeastern Montana—E. D. Cope and his team were out prospecting for new dinosaurs. Cope enjoyed the benefit of a substantial family inheritance to fund his explorations, but he was a hard worker who braved the harsh weather conditions that teams still face today in these territories. In addition, he conducted his work without highways, automobiles, ice chests, telephones, or any number of other conveniences that make field life today a comparative breeze. Although a charismatic individual, Cope had a considerable temper and tended to get on people's nerves and provoke arguments. Still, there is no denying he was incredibly productive. Not many of us today can claim Cope's 1,400 publications.

Out amid this Wild West panorama was Charles H. Sternberg, an amateur paleontologist who became a professional field prospector and collector. In 1876 Sternberg, then in his mid-20s, was living on the Kansas frontier with his young family. The entire region was rich with fossils, and these treasures piqued the interest of young Sternberg, who wanted to try his hand at mining this scientific gold rather than the metallic gold and silver that were sprouting boomtowns all over the West. A field team working for Marsh turned Sternberg down that year, but Cope encouraged Sternberg's interest in becoming a collector. The loss was Marsh's, as Sternberg would go on to become one of the most famous and well-regarded field collectors of his age, and remains a paleo-legend today.

A man of considerable imagination and passion, Charles H. Sternberg was greatly inspired by the evidence of the prehistoric worlds that he uncovered. He regarded himself as a counterpart to a big-game hunter, venturing out into the wild and relying on his experience and knowledge of the land and the "habits" of his quarry to guide him to the prizes. He called his 1909 autobiography *The Life of a Fossil Hunter* to emphasize the similarity, which is closer than one might think, even if the game hunted by a paleontologist is only skeletal and not capable of turning to the attack. As the great president of the American Museum Henry Fairfield Osborn once wrote, the fossil hunter's quest was ennobled by its unique quality: "The hunter of live game, thorough sportsman though he may be, is always bringing live animals nearer to death and extinction, whereas the fossil hunter is always seeking to bring extinct animals back to life."

Sternberg vividly pictured in his mind the creatures whose remains he so patiently removed from their enclosing strata. Considering their characteristics and likely traits in life, Sternberg would sometimes give prehistoric species evocative vernacular common names. Thus the Mesozoic *Hesperornis regalis* became to Sternberg the "Snake-Bird of the Niobrara" for its long neck and aquatic hunting habit, or "the Royal Bird of the West" for the literal translation of its name. He describes "splendid" horn cores from a "magnificent" Kansas Pleistocene bison and envisions a "noble" American mammoth. For him the *Pteranodons* were "the most perfect flying machines ever known." Clearly, here was a man who enjoyed and admired the subjects of his work. He put a passionate, capital P in Paleontology.

Not only did Sternberg make a name for himself with his superb fieldwork and numerous discoveries, he raised three sons who also became very important to the field: George, Charles M., and Levi. The specimens collected by the Sternbergs, especially their large dinosaur skeletons, can be seen in the most prominent natural history museums in America and Europe. Sternberg finds have gone to New York's American Museum of Natural History, London's Natural History Museum (formally the British Museum of Natural History), Paris's Museum of Natural History, and Frankfurt's Senckenberg Museum, among others. The Sternbergs made their name working in the American plains and Western states, but they would go on in 1912 to work in the equally productive dinosaur hunting grounds of Alberta in Canada, especially in the Red Deer River Badlands, where they followed in the footsteps of another great collector of the time, Barnum Brown, who began work there in 1910.

THE STERNBERG SONS

Trained well by their father, each of the Sternberg sons went on to accomplished careers of their own in paleontology. George is known for the great "sea serpents" (marine reptiles) he collected from the Kansas Niobrara chalk formations—now world famous for their mosasaurs, plesiosaurs, and ichthyosaurs, among other fauna. Charles M., perhaps the best known of the three Sternberg sons, developed into a paleontological analyst as well as a collector, working with the National Museum of Canada as a result of the family's early foray together into Alberta.

In addition to his collection and descriptive work, Charles went on to support the establishment of the great Dinosaur Provincial Park in Alberta, whose picturesque Badlands are today a United Nations World Heritage Site and still a very productive dinosaur-producing locality. My first visit there several years ago had me speechless both at the sheer number and diversity of dinosaur fossils and at the beauty of the area in which they are excavated. Finally, Levi Sternberg was employed by the "ROM," Toronto's Royal Ontario Museum, for which he collected many of the impressive dinosaurs that fill the galleries there. If not for Levi, the dinosaur collections would not be half as impressive as they are. Sternberg Sr. was rightly proud that the Sternbergs were considered a distinguished dynasty in dinosaur paleontology. "Thank God," he wrote, "I have raised up a race of fossil hunters."

THE STERNBERG MUMMY

Among all their most remarkable and spectacular finds, one in particular holds special relevance for us here. By 1908, the Sternberg dynasty had graduated from being mere hired hands for Cope and had earned independent status. They were able to work autonomously and provide their finds to a range of very interested institutions. That summer Sternberg led his unique family team on a "great hunt" for "the largest skull of any known vertebrate, the great three-horned dinosaur *Triceratops*." Here was a quarry worthy of this outstanding team, since barely more than a handful of good skulls existed in American museums at the time. Nonetheless, their discovery would exceed their highest expectations.

Great discoveries are often written up in the popular press with the phrase that the discoverer has "stumbled across" this or that. In paleontology this may literally be true, but it belies the great amount of systematic effort that typically leads up to the "stumble." The harder you work, the luckier you get. In this 1908 summer season, the Sternberg team searched diligently through the endless rock formations littering the Wyoming landscape of Converse County for signs of what they sought. Yet they found nothing but tantalizing weathered shapes that provoked the imagination to see in them recognizable forms where only random erosion had been at work. "Day after day hoping against hope we struggled bravely on," Sternberg wrote of the campaign. "Every night the boys gave answer to my anxious inquiry, What have you found? Nothing." Having spent several weeks each field season tramping the grounds of the West, I too have experienced the long hot days of finding nothing more than dehydration, frustration, and occasional scraps of bone. But I would not wish it to be an easy task, because hard-found fossils seem all that more precious.

The determination to stay until they earned some kind of recompense led to food stocks running dangerously low. In 1908, a convenience store was not a mere hour or two away; the nearest supply of provisions lay 65 miles distant across roadless wilderness. As the men grew hungry, Sternberg Sr. finally gave in and made the trek toward their base, but he could not forbear from searching unexplored outcrops along the way. Finally, en route, Sternberg made his hoped-for find: a *Triceratops* skull two meters long. Removing this skull from the ground occupied Sternberg so intensely that before he returned, the two boys remaining back in the field camp had run down to no food left but boiled potatoes. However, the two younger Sternbergs had a surprise for their father, which they revealed to him in the quarry where they had uncovered it in his absence. It was a duck-billed dinosaur similar to Trachodon (*Edmontosaurus*), and not only was it the most complete fossil skeleton they had ever encountered, but it was covered in places with skin impressions. It was the first dinosaur to be given the badge of "mummy."

Even in the sandstone, less photographic in preservation than rock with extremely fine grain like the Solnhofen limestone, the specific details of the hadrosaur's scales could be observed where the skin impressions were preserved. Polygonal plates—tubercles—the scale-like structures of the dinosaur's skin, interlocked in patterns around the animal's chest area. Its skin had a pebbled look, for the tubercles made repeated mosaic patterns. Similar to the architecture presented in the skin of a crocodile, or that of a bird's foot, the hadrosaur's skin showed tubercles of smaller sizes where the skin needed to be more flexible—in the areas around the joints, for example—and larger tubercles in less flexible areas such as the flank and back. Tendons could be seen along the neural spines (the tops of which you feel if you run your fingers along the back of your spine) of the vertebral column. The mummy fossil presented the clearest and most complete evidence of dinosaurian external anatomy that had ever been discovered. The animal lay in an unusual pose, on its back with the front limbs stretched upward rather than on its side, as is most common with dinosaurs of this type.

I made my first visit to the American Museum of Natural History to see this remarkable fossil, among the wealth of its displayed dinosaur remains. Cased in a box of glass, the fossil can be viewed from every angle. The contorted remains look somewhat out of place in the dinosaur gallery, given the amount of skin impressions adhering to the emaciated body of the mummy. On closer inspection the detail of the skin that has been almost "shrink-wrapped" over large portions of the body is beautiful. It is hard to believe that the skin impression was preserved in the first place.

Sternberg noted with interest the preservation of many dozens of ossified tendons crisscrossing both sides of the hadrosaur's neural spines, running diagonally in parallel rows, each the thickness of a pencil. Sternberg imagined that they must be for protection from attack by predators seeking to injure the spinal column. Today, from additional specimens, we know that the network of ossified tendons extended beyond the pelvis well down the tail. The primary interpretation is therefore that this network of ossified tendons worked to stiffen the spinal column both forward and aft of the pelvis for efficient suspension of the animal's weight over its hip, reducing unnecessary muscular strain by rendering the primary support beam for the animal's body rigid. This structure would present a typical posture with the backbone held roughly parallel to the ground, rather than closer to vertical like our own spinal column. The animal's silhouette would, in other words, probably have been closer to that of a secretary bird than a kangaroo. However, the temptation to compare extinct species with that of modern relations can lead to speculative interpretations of biology, physiology, and mechanical abilities. Alas, nothing alive today looks like a dinosaur. Dinosaurs looked and walked like dinosaurs.

The presence of these ossified tendons was a feature of the unusual degree of preservation in the specimen. Regardless of the interpretation of their primary function, their effect of stiffening the backbone is clear, and this greatly affects the parameters of how such an animal would have moved. Contrary to the old tail-dragging bipedal posture given to duck-billed dinosaurs in the decades after 1900, a hadrosaur with such a network of ossified tendons would have been physically incapable of adopting such a pose. Its vertebral column would have to be fractured to describe the arc seen in illustrations created for certain museum displays. This conclusion was not apparent in the Sternberg specimen only because its tail had not been preserved. The importance of soft-tissue preservation for correcting and clarifying our understanding of extinct species' biology cannot be overestimated.

While closer study of the bones alone might well have revealed the "fractured tail" error eventually, the extraordinary soft-tissue preservation seen in the first dinosaur mummy offered a major boost to understanding the configuration of the animal. Sternberg predicted from his observations, in 1909, that the typical hadrosaur did not carry its arms high in the air while walking, but instead more likely used its hooflike hands for quadrupedal locomotion, employing the forelimbs as arms only when standing to reach out and bring branches into convenient grazing range. In this apprehension Sternberg was ahead of his time, as scientists frequently continued to depict bipedal hadrosaurs for some decades, following the restoration produced by Charles R. Knight from this very same specimen, as directed by leading paleontologist Henry Fairfield Osborn in 1912.

For Charles H. Sternberg, the dinosaur mummy was the treasure of treasures, surpassing anything he had ever seen in his forty years of fieldwork. "Shall I ever experience such joy," he wrote, "as when I stood in the quarry for the first time?" This find, in the happy company of his sons, was "the crowning achievement of my life work!" His must have been not unlike the feelings of Howard Carter upon opening the intact tomb of the Egyptian pharaoh Tutankhamun in 1922.

The posture of the fossil hadrosaur suggested to Sternberg that the animal had died in water, which he interpreted as its frequent habitat in life. The body, he presumed, had been turned on its back by decomposition gases in the abdomen, swelling the cavity and making it buoyant. Meanwhile the heavier skull and head had angled the upper part of the body downward in the water. When the trapped abdominal gases were finally released, the body sank, the heavier skull folding sharply back and to the right shoulder as the carcass impacted the sediments at the bottom headfirst. The knees were drawn against the body. The skin caved into the abdominal area, where apparently the internal organs of the body had largely deteriorated, leaving a hollow cavity. The arms remained in the outstretched configuration given to it by the initial floating stage of the dead body, the right arm still reaching upward.

An interesting feature of some hadrosaur fossils is that they are often closely associated with some species of garfish (*Lepisosteus* sp.). The fish are often found in the gut region of articulated skeletons, but they were not eaten by the hadrosaur—almost certainly the opposite. The fish were probably scavenging the carcasses of the hadrosaurs, swimming into the body cavity to feed upon the soft tissues. Some careless fish got stuck, died, and became a fossil within a fossil. However, the fish might have also died in the same floodwaters that swept up the hadrosaur's carcass, with the two ending up together in a watery grave. There is always more than one way to interpret such associations. As Mark Twain wryly commented, "There is something fascinating about science. One gets such wholesale returns of conjecture from out of a trifling investment of fact."

Henry Fairfield Osborn published a description of the mummy find in 1912, classifying it as a *Trachodon* and noting a number of salient features, including the presence of a dorsal frill of skin projecting above the neck and backbone, a feature that could never have been discovered from skeletal evidence alone.

Osborn differed with Sternberg on the interpretation of the decaying process. Osborn proposed that the mummy must have lain exposed to the sun in order for its internal organs and flesh to have desiccated sufficiently for the skin to have been drawn so deeply into the abdomen and rib cage. He proposed that the skin had become leathery in this desiccation process, which had effectively "eviscerated" the carcass, at least as far as volume was concerned. Thus the skin survived the contraction process largely intact, wrinkling only where it no longer fit the changed geometry of the body it contained. Osborn's model for the taphonomic process theorized that after desiccation, perhaps on a sandbar, the carcass was caught in a flood that then conveyed it into the water and quickly buried it. Osborn's theory perhaps better accounts for the good condition of the skin, which one might suppose to have been more greatly deteriorated by long exposure to water than the extremely well-preserved topography evidenced in the fossil. Osborn further noted that he believed the skin to be preserved only in the form of impression casts, not actual survival of any original organic material. He presumed that the rapid burial of the stiffened and desiccated carcass promoted conditions allowing the formation of a superb mold of the body's surface details before the submersion began to decompose the soft tissues themselves.

As a clear trace fossil, the skin impressions are considered "fossilized skin," since a fossil is any physical trace attesting to the existence of past life. It is important to note that this term does *not* imply anything about the preservation of original biomolecules, which may or may not be present. Technically speaking, dinosaur mummies are a combination of both trace (skin impressions) and body (bone) fossils.

Osborn's taphonomic model for the Sternberg mummy recalls the process by which the Pompeii bodies died. They are preserved artificially as plaster of Paris casts poured into hollows discovered in the volcanic ash sediments that covered the ancient Roman city during the catastrophic eruption of Mt. Vesuvius in A.D. 79. The hollows are left by the entirely absent skin and flesh of the victims, while their skeletons remain in all but a few of these hollows. We call such remains moldic fossils. The plaster poured into these natural molds produces an artificial cast reflecting the topography of the original soft tissue, clothing, hair, and only rarely preserving actual biomolecules of hard tissue—and the skeleton—while preserving none of the soft tissue. The extreme heat of volcanic ash and pyroclastic flows that destroyed the Roman city incinerated the bodies to dust, albeit after the writhing bodies had been cast rapidly in the volcanic material.

The Sternberg mummy quickly gained renown that it has retained ever since. Henry Fairfield Osborn managed to obtain this peerless find for the American Museum of Natural History (ANMH), paying Sternberg the sum of $2,000. Known originally as a *Trachodon*, the specimen is now classified by varying authorities as an *Anatotitan* (per the American Museum) or an *Edmontosaurus*.

THE MYSTERIOUS MISSING MUMMY

Osborn noted another remarkable fact in his 1912 report: The Sternberg mummy was *not* actually the first dinosaur discovered with a substantial covering of fossilized skin. It was only the first one brought back intact. By the time of his paper, seven specimens in all had been reported with skin impressions. In 1884, Dr. J. L. Wortman had discovered the type specimen of *Trachodon mirabilis* (coded AMNH 5730), a magnificent specimen that was triumphantly mounted in the American Museum as part of its collection of Cope dinosaurs. With its huge tail intact, it is more complete than the Sternberg skeleton, which had lost its tail, hind feet, and part of its pelvis to surface erosion before it was discovered. Remarkable as the grand *Trachodon* specimen is, Dr. Wortman reported that it had originally been "surrounded by a natural cast of its epidermal impressions." In other words, it may have been an even more complete mummy than the Sternberg specimen. Confronted with the original 1884 *Trachodon* mummy, Wortman admitted that he demolished the skin impressions in order to remove just the skeleton. Surviving evidence bore out the report. "There are only three patches of epidermis remaining from the tail of this specimen," Osborn noted.

What is even more remarkable is that the Sternberg team prepared their mummy not in the controlled conditions of a laboratory, but out in the field. Nonetheless, they did an excellent job of it. Sternberg Sr. praised his son George in particular for his skillful preparation work that left the mummy skin beautifully intact, as it remains for us to see today. As soon as they noticed the presence of the extraordinary skin impressions, special care was taken to preserve every possible trace. George Sternberg achieved an outstanding level of preparation under the full heat and glare of the Wyoming summer sun, and following this work, the team brought the specimen out relatively intact. In contrast, the original dinosaur mummy was rendered into powder. We could ask for no clearer example of the value of a cautious approach and conservative preparation when faced with extraordinary finds in the field.

However, even the Sternbergs, according to Osborn, probably destroyed a good bit of their mummy's skin before realizing that it was present. While this may seem shockingly careless, it is important to realize that these skin impressions do not necessarily show up as firmer stone underlying softer, nor is there necessarily any change of color, or indeed much of anything at all to indicate an ever-so-faint impression that can be teased out of the nearly homogeneous stone by the art and skill of a master preparator.

The sediment that forms the matrix around the mummy always surrounds dinosaur bones, and the ordinary work of preparation is to pick it away so that the bones are revealed. A preparator's skill is measured at a basic level by how well they leave the bone undamaged by their attentions. This matrix may give virtually no indication of holding a revealed skin impression, until a preparator comes across a fracture plane that happens to pop off, leaving below it not an anonymous chip in the matrix below but the texture of tubercles. Suddenly you realize that a skin impression may be lurking in the stone. Where, exactly? You may only know when you find it—or when you reach bone, and realize that either it wasn't there at all, or you have inadvertently destroyed it. Even responsible collectors and preparators can unintentionally damage skin fossils. The Wortman case is only tragic because in that instance the skin fossil was knowingly destroyed, in what appears to be a case of ignorance to the importance of the find.

THE SENCKENBERG MUMMY

In 1910, the Sternbergs discovered a second hadrosaur mummy in the Lance Creek Formation in Wyoming. This one boasted a nearly complete skeleton, partial skin impressions, and the preservation of the horny beak sported by the dinosaur, a trait it shared in common with many other herbivorous ornithischian dinosaurs.

The Sternbergs labored hard for several months to excavate and transport the specimen, which weighed close to 10,000 pounds. After excavation, the mummy was taken from its remote location to railheads for its journey across the United States and overseas. This specimen was sold to the Senckenberg Museum in Frankfurt, Germany, where it remains on display today.

Much of the partial skin impression envelope originally surrounding this dinosaur mummy has been removed through preparation. The most interesting body part for me is the exquisitely preserved hands, complete with partial mittens of skin. The skeleton that supports many patches of skin impressions over the neck and back is relatively 3-D, in so far as the bones have not collapsed flat, as with so many fossils. In many respects, this is a not-so-well-preserved twin of the AMNH mummy. Charles H. Sternberg speculated in his memoir *Hunting Dinosaurs in the Bad Lands of the Red Deer River, Alberta, Canada* that "the dinosaur died in quicksand, which helped to preserve its contorted death pose as it fought to escape." Rapid burial is almost certainly a key factor in such preservation, but this interpretation cannot be confirmed by the records of the site's sediments.

BARNUM BROWN'S CORYTHOSAUR MUMMY

In 1912, Sternberg's great peer in the world of Western dinosaur collection discovered his own dinosaur mummy, a largely complete specimen with substantial soft-tissue preservation and skin impressions (AMNH 5240). The peer was Barnum Brown, and the dinosaur—a new genus—was *Corythosaurus*. This was another hadrosaur, similar in form to *Edmontosaurus* and *Anatotitan*, but with a different "crown" structure on its head. In 1914 Brown discovered a second specimen of this dinosaur, a nearly complete skeleton (AMNH 5338). Between the two he was able to publish a superb description of *Corythosaurus*, and the composite analysis benefited greatly from the soft-tissue preservation seen in the mummy.

For the first time the hadrosaurs' ossified tendon network was completely preserved, showing that it did run from the back well down the tail, creating the stiffened backbone that would not have supported the upright vertical posture reconstructed by Osborn in 1912. The superior preservation even showed that there were two superimposed layers of tendons running diagonally opposite each other, creating a reticulate, trellis-like pattern.

Muscle tissue traces were fossilized in the *Corythosaurus*, visible as patterns indicating the ischio-caudals (linking the base of the tail to the anchorage of the pelvis) and what appear to be the sphincter muscles in the area of the cloaca, almost a fossilized derrière! This was an astonishing degree of preservation—to be able to examine a dinosaur's musculature with the naked eye was unprecedented.

Brown's *Corythosaurus* also showed a distinct dorsal frill, a vertical flap structure running down the back. The skin impressions showed tubercles, but lacking any pattern. On *Corythosaurus*, the tubercles showed the scaling—smaller in flexible areas—but not the elaborate patterning that had appeared on Sternberg's mummies. Rows of conic tubercles, however, could be seen in the belly area.

Brown observed that skin impressions were so commonly associated with articulated specimens of hadrosaurs that it was actually exceptional *not* to find all or part of the skeleton surrounded by a skin impression envelope. Osborn hoped that the spectacular mummy specimens on display would heighten awareness of this possibility and prevent the accidental destruction of any future dinosaur mummies. Whether or not other mummies were inadvertently destroyed or never recognized, though, no more dinosaur mummies were reported for many decades.

A TITANIC LOSS TO PALEONTOLOGY

After publishing his memoir, Charles Sternberg went on to find not only the 1910 Senckenberg mummy but still more hadrosaur specimens after that, including another particularly fine one with skin impressions. The British Museum of Natural History purchased many of these hadrosaur specimens, along with a large collection of other fossils from Alberta. All these were shipped across the Atlantic in 1916, in the midst of World War I. The Canadian transport vessel *Mount Temple,* carrying the 22 boxes loaded with fossil bones, was targeted by a German raider ship painted to resemble a tramp freighter, so that it could sneak up on merchant ships and then destroy them. Fitted with camouflaged deck guns and torpedo tubes, the SMS *Moewe* was heavily armed. After a brief attack the *Mount Temple* was sunk with her fossil cargo. The vessel lies today in more than 14,000 feet of water, in the general region of the wreck of the *Titanic.* The *Mount Temple* is the only known sunken ship to carry major dinosaur fossils. In his later memoirs Sternberg was very bitter about this loss of the irreplaceable treasures his hard work had produced.

The manifest of fossils aboard the *Mount Temple* does not clearly list what was shipped, and while the figure of "two hadrosaurs" is often quoted as the primary loss, Alberta paleontology technician Darren Tanke has researched the matter and concluded that as many as four hadrosaur specimens were lost, although probably none of them were complete. There does not appear to have been "two mummies," as is commonly reported.

LEONARDO DINOSAUR MUMMY

On July 20, 2000, a remarkable discovery came to light in the Judith River Formation in Montana. It dated back 77 million years, to the late Cretaceous Period. Discovered by staff and volunteers from the Judith River Foundation (JRF), the find was a subadult hadrosaur, a *Brachylophosaurus canadensis.* This was another of the so-called duck-billed dinosaurs, a plant eater about seven meters in length. Excavation would reveal that the skeleton was nearly complete and fully articulated. That means the bones remain after fossilization largely in the same relation to each other as they did in the animal's life.

Such articulated finds rightly gain media coverage and frequently command pride of place in museum displays, but they are not what paleontologists normally find in the field. Even partial or disarticulated specimens are rare. Isolated bones are the norm—a sampling of bones from a carcass that was largely devoured, torn apart, and scattered by predators and scavengers before final burial—or isolated bones that turn up here and there, having been carried along prehistoric rivers or by flood washes, leaving puzzle pieces that today are collected piecemeal. The find of even a single articulated limb can be a valuable discovery. More than half of an articulated dinosaur—either half!—is truly an occasion for celebration.

Excavation of the Judith River Formation discovery was carried out the following year, beginning in May 2001. Overburden above the skeleton was so extensive and resistant as to require a bulldozer, drilling, and blasting. When only a meter of sediment remained above the specimen, the team took up hand tools and began working their way down, employing pneumatic tools where the sandstone grew tough and concreted. As the find was gradually exposed, project leader and JRF paleontologist Nate L. Murphy and his team found that their *Brachylophosaurus* appeared to be not only articulated but almost complete, with only a section of its tail missing. Much more striking than this degree of completeness, however, was the fact that diggers kept uncovering what appeared to be skin impressions.

Ordinarily a dig team separates a fossil dinosaur skeleton into manageable portions for transport out of the field. Many bones may be individually isolated before being protected by plaster jackets and carried away. Smaller bones may be removed in groups for convenience. The separation of elements creates no special problems because the original orientations are all recorded carefully before anything is taken out of the ground. In the case of encountering fossil skin, however, one does not want to just chop through it in order to isolate separate pieces of the skeleton beneath. Fossil skin impressions are so precious that one wants to protect every square inch of them and sacrifice as little as possible. As they dug downward, Murphy's team kept hitting skin impressions here, there, and everywhere on their *Brachylophosaurus*. To preserve whatever traces might remain, they decided the whole beast had to be taken out of the ground in a single block. Only an unconnected tail sequence was removed separately.

To surround the large block in a protective field jacket required approximately a ton of plaster. To give strength to the package, it was surrounded by a sturdy steel framework of heavy square tubing. Altogether, the single mega-jacket weighed 6.5 tons. As we will discover later, Tyler's fossil gave us an even larger plaster of Paris bill!

As has become the custom ever since a *Tyrannosaurus rex* fossil was named Sue, the striking new find was soon given a name: Leonardo. The name was derived from that of two lovers who had carved their names in the rock face above the excavation site many years ago. Who said that paleontologists are not romantics? The fossil's soft-tissue preservation ratcheted up interest dramatically as lab preparation proceeded at the Judith River Institute. Leonardo appeared to be covered to a substantial degree with fossil skin. Murphy reported that it was present over an astonishing 90 percent of the preserved body, lacking only on part of the right forelimb and the right side of the skull. This degree of preservation indicated the presence of a feature that could not be detected from bones alone: a vertical comb or frill of skin that rose along the spine from the skull to the middle of the body. Whether the fossil skin represented any preservation of original biomolecules, or whether it might simply be impressions made by the original skin preserved as mold-and-cast and preserved as a trace fossil, the investigators have refrained from concluding either way. They have adopted the term "integument trace" to refer without commitment to what the popular press would call "fossil skin" or "skin impressions." Having seen many images of the fossil and spoken to Murphy at length, I can confidently state that Leonardo is a remarkable find. The key to all such finds is that the data be used in conjunction with other fossil finds, both old and new, to help fill gaps in our knowledge.

As reconstructed by Murphy and his colleagues, the body of their subadult *Brachylophosaurus* floated down a river headfirst until it hit a sandbar with its neck and right shoulder. Somehow the bulk of the animal's flesh appears to have been preserved in the fossil in the area where it initially impacted the sandbar. Most of the rest of the body appears to have lost its flesh before fossilization, with the integument trace ending up tightly wrapped against the skeleton like the skin of a desiccated mummy. In the shoulder area, however, the fossil appears to reflect the geometry of the musculature of the living hadrosaur. The investigative team interprets the sedimentology and the preservation as indicating rapid deposition that covered over the carcass before it could be scavenged. What exactly accounts for the elimination of the internal flesh, and yet the preservation of the skin, in what was clearly a wet environment, which would ordinarily promote rapid and thorough decay, is not yet clear.

During preparation of Leonardo, a hole was accidentally made through the skin in the belly area. This hole revealed a darker mass within the area enclosed by the animal's ribs: a "lignaceous mass" of what appear to be gastric tract contents in the form of numerous chewed-up and partly digested plant fragments. These are reported as only 1-5 millimeters in size, presumably reflecting effective mastication by the hadrosaur's powerful tooth batteries. Because the mass of plant material appeared to be completely contained within continuous soft-tissue traces, the Leonardo investigation team concluded that it truly represents the gastric contents of the dinosaur, rather than the remains of plants that might have randomly collected in the corpse after the animal died. The Sternberg hadrosaur mummies had not offered the same kind of evidence in the stomach cavity, since their body cavities appeared to have been incompletely preserved. The investigative team for Leonardo concluded from their examination that the stomach cavity of the new mummy was, quite remarkably, still intact. This is a surprising state of preservation because under ordinary conditions an animal's stomach acids and other digestive juices begin acting to break down the animal's own gastric tissues very soon after it dies. The depositional conditions that would account for the apparently excellent condition of the digestive tract in Leonardo remain to be identified and explained.

A second hole was also inadvertently opened in the Leonardo fossil during preparation, this one in the pelvic area. The second opening revealed additional lignaceous material within the body, presumably from farther down the digestive tract. As one would anticipate with more fully digested plant material, the material here is more homogenous, and the fragments are smaller and much less identifiable. Its compressed aspect is consistent with the state of meal remains more completely processed by an animal's system.

GUT CONTENT ANALYSIS

Murphy and the Leonardo investigation team presumed that the heavily chewed-up and partially digested plant material in their hadrosaur's gut would not likely furnish evidence complete enough to identify the species of the plants. The team decided not to attempt structural analysis of the plant remains, but instead to seek analysis of the pollen suite contained in the presumed gut contents. The pollen removed from this lignaceous mass would not be restricted to only the plants deliberately eaten by the hadrosaur when it was alive, so the suite would not produce a list of probable hadrosaur diet species, but it might provide a sense of the paleoecology of the environment in which this animal had lived and died.

Pollen spores are marvelous paleoecological evidence, because they survive the ages so well, and because via microscopic study the spores can be traced to plant types frequently down to the specific level. Pollen in a given environment often scatters and mixes through local breezes, creating a signature that can be read by palynological (the science of pollen) analysis. So important is this type of study to paleontological interpretation that palynological laboratories now offer their services on a freelance basis.

Analysis of the spores identified 40 distinct species of plants, according to analyst Dennis Braman of the Royal Tyrrell Museum in Drumheller, Alberta. The suite indicates a warm, moist subtropical environment conducive to the growth of ferns and even liverworts, which require continuously humid conditions. Thus we must picture this *Brachylophosaurus* living in an environment ideal for promoting rapid decomposition, adding to the puzzle of the carcass's excellent preservation.

The gut contents of Leonardo also provided a hitherto unique opportunity to examine the material for traces of parasites. Researcher Karen Chin of the University of Colorado at Boulder, a specialist in the analysis of prehistoric trace fossils, examined 17 samples from Leonardo's gut contents. With colleague Justin Tweet, Chin identified 200 suspected burrows of tiny parasites, but they all appear to be of one kind rather than the suite of scavengers one normally finds in a decomposing body. The researchers theorize that these parasites were already living within the body of the dinosaur before its death, and that the burrows represent activity subsequent to death, during which time the body was somehow isolated from attack by other parasites that would have caused normal deterioration of the body. Chin's analysis provided the first evidence for parasites in the gut contents of a dinosaur.

X-RAY INVESTIGATION

The exceptional preservation of a mummified dinosaur invites extreme curiosity about its internal structures, but responsible investigators are loath to bore destructively through the precious skin fossils in order to explore what might or might not lie beneath. Accordingly, advanced technology is now being brought to bear on well-preserved fossil specimens, in experimental approaches via digital imaging. In the summer of 2006, Leonardo was examined by a Kodak Industrex ACR-2000i digital x-ray imaging system, which was brought to the JRF location in Malta, Montana. The x-ray analysis reportedly produced "more data than any of the paleontologists anticipated," but the results have not yet been published in full.

The technique used by Murphy and his team will help unlock some of the secrets of Leonardo, but to truly "look inside" the fossil would require the use of an industrial CT scanner. Such scanners can slice through metal engine blocks to reveal their inner components in a noninvasive way. Different elements of the engine or fossil can also be extracted using powerful software to image where the eye was never meant to see. Early in our research of Tyler's fossil, we realized that such insight would be invaluable, so we started our search for such a suitable CT machine.

If Leonardo perished in a warm, moist environment of ferns and liverworts, how could its body have been mummified rather than decaying rapidly? Murphy and his colleagues propose that some unusual bacterio-chemical processes must have broken down and liquefied the flesh and guts of Leonardo, while somehow leaving the skin preserved. However, this scenario does not explain the specific nature of the process, or how it produced this most unusual find.

It is perhaps fitting that so remarkable a find as Leonardo should retain some of his grave secrets for the time being. The popular press crowned him king of the mummified dinosaurs and called him the best-preserved dinosaur in the world, for if the reports are accurate, the soft-tissue preservation of Leonardo, together with its apparent gut contents, give it an edge over even the Sternberg mummies.

Such was the state of affairs when Tyler encountered his own remarkable find in the American West. Here was a unique opportunity to approach a prehistoric find with 21st-century methods and technology. What would follow was a series of bizarre coincidences that eventually brought my and Tyler's paths to cross. For collaboration to be successful, first serendipity has to play an important part.

CHAPTER FIVE
MANCHESTER IN THE BADLANDS

"I have never let my schooling interfere with my education."
—Mark Twain

FOSSIL HUNTING HAS BEEN PART of my life for as long as I can remember. My childhood was spent searching the Jurassic rocks of Somerset for the fossil shells that were strewn among the dry-stone walls and cairns of my garden. I soon had a significant pile of fossils attracting large amounts of dust in my parents' house. The simple wonder of touching the fossilized remains of a once-living organism has never left me. When we pick up a fossil and scrutinize its shape and completeness, our mind's eye reconstructs the whole or part of the organism preserved. The abstract coils of a shell or convoluted processes of a vertebra become part of a living, functioning animal. Paleontology relies heavily upon an individual to know as many body parts as possible, both ancient and living, to help determine if these abstract shapes actually mean something. This is where an interest in the arts, especially sculpture, has helped me a great deal in interpreting fossil remains.

My father worked in the print industry, and my mother kept my sister, brother, and me in order at home. My father was a fairly strict soul, who has mellowed in recent years, but I recall him working long hours when I was a child. My mother, who

I might proudly add is Danish, put up with my older brother's hard rock obsession of the Led Zeppelin variety and my rock obsession of the landscape variety.

When at school, I was not the brightest spark in the class. The intellectual constraints dictated by the curriculum did not suit me. My overall school report could quite adequately be summed up as "could do better." This was not from lack of trying, but lessons were a laborious chore that provided me with little to no stimulation. I was not able to study geology until I was 16 years old. Then I discovered art.

Many of us are fortunate to have teachers that helped mold our minds and futures. Luckily for me, Mary Price was teaching three-dimensional art (sculpture and pottery) when I passed through the hallowed halls of Well's Blue School in Somerset. Miss Price was a patient soul who could look straight through you and see what you were about. When I was first handed a lump of clay to sculpt, it was a wonderful experience. I could create a three-dimensional shape that I wanted to create, not what was dictated by others. After a while I became more familiar with the materials, both the clay and the tools used to shape my mind's creation. Miss Price gently steered me from sculpting mushrooms to birds and eventually the human form. The most powerful tool available was not the clay used to sculpt or the knives, spatula, and wires to form, but my eyes. The simple art of observation was one of the keys to learning and understanding three-dimensional form, a trait I wish more students would apply when viewing a fossil for the first time. Often the answer they seek lies in simple observation, appraisal, and interpretation.

At the end of the many public lectures that I give in museums, schools, and universities, parents and students often approach me to ask which is the best route into paleontology. My answer is one that they do not expect. I look at the son or daughter and ask, "What is your favorite subject at school?" "What really interests you?" At first both parent and child look at me confused, because they were expecting a clear prescription of what to do and where to do it. Yet paleontology has many facets that can be approached by a multitude of routes, both academic and nonacademic. Some of the most brilliant paleontologists and fossil hounds have never seen the insides of a lecture hall. The modern science of paleontology draws knowledge and applications from fields as diverse as materials science to computer science and from biology to mechanical engineering. Each discipline offers a new way of solving old problems, often with positive outcomes. The ability to visualize a complex three-dimensional form when only partially exposed is where my sculpture background often helps in the field. If you place a chisel in the wrong place when excavating a vertebra and drive it home, you are sometimes met by the bone-crunching ring that sends shivers down your spine—but doing a lot more damage to the spine of the entombed animal. The ability to mentally project the form of the bone into the rock is something that years of experience and a background in sculpture have helped me do. I have thankfully not heard the crunch of bone at the end of a chisel for many years.

I joined my older brother, Steven, at Nene College in Northampton, then an associate college of the University of Leicester. I studied the closest that the United Kingdom had to offer in terms of a liberal arts degree, with a minor in art and drama and my major in earth sciences. The most recurrent part of my studies comprised meteorology and applied climatology, with occasional forays into fossils and evolution. The one thing I retain from my time at Nene is a typically British obsession with the weather.

My final-year dissertation focused on the paleontology of Whitecliff Bay on the Isle of Wight. While working on the island, I visited the local Geology Museum in Sandown to talk rocks and fossils with the curator, Steve Hutt. After graduating I spent several months working in the print industry, but my heart was not in the work. Then my parents got a phone call out of the blue from Hutt, asking them to tell me about a job coming up at the Museum of Isle of Wight. Luckily, my mum was in that day, she took the call, and very soon I was traveling to the Isle of Wight as the curatorial assistant at the museum. This was to be the first of many postings in U.K. museums. I had started my career as an employed paleontologist, a position I now know is a rarity in its own right.

The Isle of Wight is a paleontological paradise. Lying just off the south coast of England, the diamond-shaped island is split geologically in two, with Cretaceous rocks to the south and Tertiary rocks to the north. The Cretaceous chalk downs form an east-west spine to the island. The oldest rocks on the island are roughly 125 million years old. These rocks, locally called the Wessex Formation, outcrop on the southwest and southeast coasts. These two locales were to provide me with the most useful fieldwork of my early career. Often with the experienced guidance of Hutt and the many amateur collectors who trawled the coast, I started hunting for dinosaur fossils. I couldn't quite believe that I was being paid a salary to do this work.

The late 1980s and early '90s were good years for hunting dinosaurs on the Isle of Wight. I cut my teeth working on dig sites of the new predatory dinosaur *Neovenator*. On a collecting trip to the southwest coast with Hutt, he discovered a huge sauropod below the Barnes High Sandstone. The most common dinosaur was *Iguanodon*, or "pigs" as many of the local collectors referred to them. A winter southwesterly gale in the English Channel, coupled with spring tides, often led to isolated dinosaur bones being strewn across the beaches of the island. One day wading off Yaverland beach, on the southeast coast, I filled two rucksacks strapped to both my back and front to the brim with dinosaur bones. Since the advent of *Jurassic Park*, however, those days are gone. Far more collectors work the coast of the island, picking it clean of bones. If you are lucky today, you might pick up the odd rolled piece of bone, gently smoothed by wave action.

While working at the museum I discovered one of the most enjoyable elements of my work: teaching. Leading school and college groups to hunt for ammonites off St. Catherine's Point was simply terrific fun. Watching faces light up as they discovered their first fossil was gratifying, and to be able to explain what they were looking at made the trips even more worthwhile. That's because an exciting find discovered out of context, while aesthetically pleasing, has little depth. When a fossil's position in geological time, sedimentary environment, and ecological niche is properly explained, a vast dimension can be opened to extinct lives and forgotten worlds. Small hands grasp their hard-found fossils much tighter as knowledge of their discoveries becomes clearer. A trip to the seashore can become an adventure to another world. Sharing with young and old the story of prehistoric life continues to be the most rewarding part of my life.

Working with dinosaurs, you get used to folks stuffing cameras or microphones into your face and asking you to speak in sentences. The undergraduate courses in drama I took suddenly became useful. A call came into the museum one day to see if a "geologist" could help a television production explain coastal erosion for a school program. I was volunteered and despatched to Whitecliff Bay. Waiting for me at the holiday camp above the bay were a TV crew and a class of schoolchildren from an inner London school. As we staggered down the cliff path toward the beach, the director of the program explained how the school kids were going to rush excitedly toward the sea. As the crew hit the beach, though, the kids started sprinting for the surf line. In moments they were already waist-deep and still fully clothed in the sea—with red-faced teachers screaming for them to come back. Not the most successful start with television. However, I soon realized that the story of life on earth was one that was worth telling to as wide an audience as possible, and television offered such an audience.

The island provided an irreplaceable period of time in which to become certain about my career path. What had been a childhood hobby was turning into a way of life. To make a way of life an occupation, I needed to go back to college. I felt that an undergraduate degree was not going to get me to where I wanted to be. Where that was, I was a little unsure, but all whom I spoke to advised that I needed at least my master's degree.

One weekend up in Manchester, visiting the parents of my girlfriend, I was flicking through the *Manchester Evening News* jobs section when an advertisement caught my eye. A paleontology technician was wanted in the Department of Geology at the University of Manchester. The next paleontological chapter in my life was about to start. At the interview for the job it was suggested that I undertake my master's while working on the research project as technician. Fortunately, the project was funded for three years, giving me plenty of time to hunt for the world's oldest land animals in Silurian rocks from the Welsh borderlands.

My job at Manchester involved drowning vast quantities of 414-million-year-old shale in concentrated hydrofluoric acid. Nasty stuff. Hydrofluoric acid is one of the single most unpleasant acids, not as if there are any pleasant ones, that you can work with. The HF molecule is small and can pass through your skin painlessly, but when it hits bone, you feel like someone is beating you with a very large hammer. This I know, having had a tiny drop penetrate several layers of the protective glove I was wearing make contact with my index finder. I did not repeat the exercise.

I spent months macerating the tough shale we collected from various localities around the Welsh borderlands. Specifically targeting organic-rich horizons, we built a vast microscopic collection of plant and arthropod cuticles from a unique window in time that had preserved fragments of one of the world's earliest land ecosystems. As we saw in an earlier chapter, the organic remains entombed in the rocks we studied had been washed off an ancient landmass during immense storms into a shallow marine environment. As the seasonal storms dragged over the shallow seas, they caused deepwater sedimentary features to form, locking these turbulent times into the sedimentary record.

The fauna was dominated by marine arthropods, and these eurypterids were to form the subject of my master's thesis.

The site did contain the occasional monsters. More than six feet long, coupled with powerful claws and complex multi-toothed mouth parts, were top predators at a time when our distant vertebrate fishy ancestors were beginning to expand. Some have suggested the early radiation of vertebrates was a function of competition with the eurypterids.

Working in the paleo-lab alongside me were other postgrads who were all firm arthropod researchers. Jason Dunlop (now at the Humboldt, Berlin), Lyall Anderson (now at the University of Cambridge), and Simon Braddy (now at the University of Bristol) were all studying for their Ph.D.'s. Dunlop's main impact on my time in Manchester was to help shake off my arachnophobia. Jason had a vast collection of various species of theraphosid spiders (tarantulas). One species, the Goliath bird-eating spider (*Theraphosa blondi*), I swear had clogs at the end of each hairy leg. Each time Jason would let his pets out for a quick run, you could hear the fine tapping and tumbling footfalls making their way toward you along the melanin bench tops. I soon had to get used to these giant, eight-legged fur balls with their 12-inch leg span.

My work on eurypterids was regularly punctuated by running field trips for the undergraduates and postgraduates to the Isle of Wight and Yorkshire Coast, hunting for dinosaurs and marine reptiles respectively. The draw of vertebrate paleontology was still strong, but I had to finish my master's before I could pursue more bony questions. As my contract at Manchester drew to a close, I started applying for new positions and fortunately the first I applied for I got, meaning I had to finish my master's posthaste.

I submitted my master's thesis on the paleoecology of eurypterids the week that I started at Clitheroe Castle Museum in Lancashire. Perched on a Carboniferous reef knoll, the castle and museum look down on the picturesque town of Clitheroe. I was both nervous and excited, as this was my first curator position in a museum. I inherited an eclectic collection of weaving looms,

World War II gas masks, and thousands of Carboniferous fossils from the local area. Not a dinosaur in sight, but that was about to change.

At the job interview I foolishly wagered that I would double the museum's best-ever visitor figures during my first year. I intended to do this with a simple device, a dinosaur exhibition: "Dinosaurs Come in from the Park!" Fresh on the tail of *Jurassic Park*, I decided that Lancashire needed an injection of dinosaurs into the Carboniferous. An ex-Manchester Ph.D., Andy Jeram, kindly offered a stack of dinosaur material to display from the Royal Ulster Museum, where he worked. I hotfooted it over to Belfast, while cases were built for the exhibition back in Clitheroe. The display area was small, but we managed to stuff it with dinosaurs bones from Mongolia to North America, not to mention material from the azure shore of the Isle of Wight. By the end of the exhibition's run, we succeeded in hitting our visitor target.

That same summer I spotted an advertisement for a Ph.D. at the University of Sheffield on dinosaurs. Tempting. I now realize that Ph.D.'s are not normally advertised in national papers, so again I was just plain lucky picking up the paper that day. The Sheffield project would study the Middle Jurassic dinosaur tracks of the Yorkshire coast. Undertaking a Ph.D. is a big commitment for anyone, and in the United Kingdom it entails three years of research, usually followed with a year's write-up. I was fortunate that I received a scholarship from the university to pay my fees and provide a small stipend. My supervisors, Mike Romano and Martin Whyte, had been exploring the dinosaur tracks of the Yorkshire coast for many years, and wanted a post-grad student to help unlock some of the secrets of track formation and preservation. I was keen to take up the challenge and was soon burying myself in track literature, fieldwork, and fossils.

My Ph.D. tenure was a rather tough ride in terms of finances and family. My wife, whom I had married while working at Manchester, had kindly offered to help support me during my Ph.D. We planned that my wife would take a break after I completed my studies and got a job. The best-laid plans of mice and men! The day I handed in my notice at Clitheroe Castle Museum, I returned home and was greeted with the news that I was to be a father. Alice was born toward the end of my first year of study; Kate was born the following year. The financial hardship of starting a family with only minimal income cannot be understated. I worked all day while continuing my studies, mostly at night. I am sure my supervisors bore as much stress as I did, and I am still grateful for their patience.

My work on the dinosaur tracks of the Yorkshire coast soon extended to my old fossil-hunting grounds on the Isle of Wight. I was also exploring the Yorkshire coast for marine reptile fossils from the marine succession underpinning the dinosaur track-bearing horizons. In the first year of my Ph.D. I bagged a 185-million-year-old marine crocodile from Whitby. The skull and much of the skeleton still remained. By the end of my Ph.D. we had found plesiosaurs, ichthyosaurs, and the first (and oldest) sauropod dinosaur from the Yorkshire coast. Dinosaur tracks were also coming thick and fast. The more work I undertook in the lab modeling track formation, the more tracks I began to recognize in the field. The complex three-dimensional print that occurs beneath the sole of a foot displaces more than just surface sediment. The sediment below the foot also shifts, compacted by the force. The combined 3-D form of a surface and subsurface track has the potential to convey information on the track maker's foot morphology, speed, gait, and step-cycle kinematics.

As I slowly tracked my way into my fourth year and the big Ph.D. write-up phase, a job came up at the Yorkshire Museum, a gem located in the beautiful city of York. Still in the throes of writing the dissertation, I applied and was lucky, again. Yet working full-time while still writing a thesis was not such a good idea.

At York I was appointed to the position of Keeper of Geology, a job that I had often hankered after while undertaking my research on the Yorkshire coast for my Ph.D. York is the quintessential walled medieval city, with ample portions of Anglo-Saxon, Viking, and

Roman archaeology found in the city. Working there was quite an experience, given the museum had a collegial approach to almost every facet of its work. In other words, I gained a vast amount of experience in everything from public lectures, exhibition design and building, touring exhibitions, and IT infrastructure roll-outs. I even dressed up as a Saxon warrior once. Life was never dull in York.

The highlight of my time there was possibly the visit of Queen Elizabeth II and the Duke of Edinburgh to unveil a new bell for the Archbishop of York. Their visit also officially opened our "Walking with Dinosaurs" exhibition that we had built for the BBC. It was great fun introducing the King of the Tyrant Reptiles (*T. rex*) to the Queen. The royal family's involvement in paleontology had a long history, for Queen Victoria's husband, Prince Albert, was instrumental in building the Natural History Museum in London. The opening day was a great success, but I could not help thinking that the Queen must think the antediluvian world smelled of wet paint! Almost every inch of the museum that the royal party viewed was freshly painted a few days before their visit.

While at York I also took up a part-time lectureship teaching vertebrate paleontology at the University of Liverpool. Once a week I would rise at five a.m., make my way to York railway station, and travel across the Pennine hills to Liverpool. I would arrive just before nine, giving me time to sprint up the hill from the station to the university to teach my course. It was a long but worthwhile day. My experience in Liverpool made me decide that I needed an academic position, as I enjoyed the teaching and research components that were formally absent in museum work. After five years in York, it was time to move on.

The curator of paleontology position came up at the Manchester Museum at the University of Manchester. Once again I applied and once again I was lucky. My idyllic life in York was about to be transformed into working in the heart of a big city in the U.K.'s largest university. The director of the museum was Tristram Besterman, who was a legend in U.K. museums. Besterman's key interests were, and continue to be, museum ethics, scholarship, management, and leadership, often focusing on the social interaction and value, accountability, and sustainability of museums. He is particularly interested in the social purpose of museums as democratic places of social engagement and the way that, at their best, they startle, challenge, and delight through the revelation of narrative embedded in objects and contexts. This led to numerous fruitful exchanges of ideas and actions; among them was the acquisition and installation of a life-size skeletal mount of a *T. rex* in full sprint. Besterman approved the acquisition, based upon the potential public engagement and scientific outcomes that I proposed in a business plan. Basically, I got a big dinosaur to play with. That made for one happy curator.

After two years in that post, I attained the Museum Joint Academic appointment, where you find me today. In the museum I am a research fellow, and in the School of Earth, Atmospheric, and Environmental Sciences, I am a lecturer in paleontology. Both jobs keep me very busy serving two taskmasters.

I was in my office in the University of Manchester, working with one of my master's students, Emma Schachner, when I first heard about a dinosaur that would occupy so much of my time in the future. Schachner had chosen to work on the ornithopod dinosaur *Tenontosaurus*, of which we had a composite skeleton from the Cloverly Formation in Montana, held in the museum's collections. She was having fun trying to figure out the particular articulation of the forearm. As I helped her reposition the heavy dinosaur bones, Schachner spoke about a dig she had been involved with that summer. Schachner had been working with the Marmarth Research Foundation on a hadrosaur excavation. When she mentioned a hadrosaur with skin impressions, I had to learn more about what sounded to be an incredible find. After several e-mail exchanges with a young man called Tyler Lyson, I agreed it would be excellent if we could meet and talk dinosaur.

Tyler had quite a story to tell me. He had grown up in the Badlands of North Dakota, prime territory for dinosaur prospectors. One of its richest areas is the Hell Creek Formation, a large "package" of mudstones and sandstones, deposited 67-65 million years ago and now forming much of the Badlands landscape in South Dakota, North Dakota, and Montana. The Lance Creek Formation in Wyoming is equivalent in age, but is sandier than the typical mudstone-dominated environment at the Hell Creek. Many argue that they are one and the same. The Scollard and Frenchman Formations in Canada are also similar in age and environment, but modern geography has had an impact upon the naming of these rock units. The top of the Hell Creek Formation is marked by the K/T (Cretaceous/Tertiary) boundary, which marks the point in time where the most famous mass extinction occurred, marking the end of the Age of the Dinosaurs. The K/T extinction is, however, the smallest of the "big five" extinctions that have been studied. The boundary between the Permian (290-251 million years ago) and Triassic (251-204 million years ago) Periods marks the point when life was almost extinguished on the planet. At the P/T boundary an estimated 95 percent of all marine life and 70 percent of all land families became extinct. The K/T extinction wiped out 70 percent of the species.

Although *Edmontosaurus* appears abundantly among the Hell Creek Formation dinosaur fauna, it shared a prehistoric landscape with many iconic dinosaurs, including *T. rex* and *Triceratops.* A huge amount of excavation has been undertaken on the Hell Creek Formation, possibly because of *T. rex*. While many areas of paleontology complain about the disproportionate time the media gives to dinosaurs, paleontologists who study dinosaurs in turn complain about the time devoted to this one predator. The Hell Creek Formation has many other species of dinosaur, reptile, bird, mammal, amphibian, insect, and plant that combine to give a snapshot of life in the late Cretaceous of the American West.

Tyler's startling fossil of the *Edmontosaurus* started to be excavated when he and the team from the Marmarth Research Foundation (MRF) were busy excavating one of the *Triceratops* sites that they had found. Set up by Tyler and his family, the MRF is a nonprofit organization that enables anyone with a passion for paleontology to share their excitement in the Badlands of North Dakota. The primary objective of the MRF is to locate, excavate, and prepare fossils from the Hell Creek Formation in order to learn new scientific information and create a collection to display in a future museum. The half dozen or so people on the MRF staff all have extensive experience both in the field and with fossil preparation in the laboratory. Small fees raised by people attending each excavation funds the MRF field, laboratory, and research programs.

Several organizations around the world give professional paleontologists and amateurs the opportunity to excavate anything from mammoths to dinosaurs. Such programs help bring home the difficulties of excavating the past. Although an individual might first expect the downdraft of a helicopter's rotor blades to gently waft the sands of time away from their articulated skeleton, the truth soon hits hard. Vast amounts of backbreaking work often have to be done before a single shard of bone appears from the unyielding earth. As volunteers graduate to diggers and then to excavators and eventually to working on the actual bones, a practical approach to learning that has been working since the beginning of the 19th century is transferred to another generation.

An excavation site soon looks like a hive of activity, with heads bobbing up and down and banter keeping spirits high, akin to a prairie dog town. At times I wonder what prairie dogs think of their strange neighbors digging through the same ground in which they live. The wheelbarrows, shovels, picks, brushes, and burlap soon litter the site, all being applied in the simple pursuit of fossils. Fortunately for many of the MRF volunteers, they have merely a one- or two-week stint digging in the Badlands. Paleontologists, like prairie dogs, are pretty much sentenced for life.

The hadrosaur site, originally found in 1999 while Tyler was still in high school, was still protected by many tons of Hell Creek sediments, so priority had to be given to the sites where overburden had been removed and bones exposed. The MRF *Triceratops* site proved a little disappointing, with the bones diminishing faster than the team's enthusiasm. Faced with the diminishing returns, Tyler took the opportunity to open a new site. First he looked through his field notebooks for a suitable one, and came across his note on the hadrosaur site. Armed with the GPS coordinates, it was a simple matter for Tyler to relocate the site.

The site was as he had left it five years earlier. Dinosaur sites are not known to move by themselves, but before the advent of GPS you could easily believe they did. Relocating a fossil site in the vast expanse of the North Dakota Badlands is no mean feat. Once in place, the initial dig was planned, resourced, and executed, and a team of MRF volunteers began by clearing tons of dirt and rock. Most people are surprised at the distance that excavators work from a fossil they are digging up. Digging too close to a large dinosaur skeleton can easily end with a team member sticking a pick into a bone just below the surface. So in addition to patience, caution is also a useful asset for a field paleontologist and excavation team.

When the excavation started, Tyler had thought he might be lucky enough to find a tail or even a partial skeleton of a hadrosaur. Like any dinosaur find, the more he uncovered, the greater potential the specimen had to unlock prehistoric secrets. While hadrosaur bones are not rare in the Hell Creek Formation, any articulated skeleton of a dinosaur would be a prize. Putting this into context, it is worth remembering that the most famous Hell Creek dinosaur, *Tyrannosaurus rex*, is known mainly from many partial skeletons; only two have more than 50 percent of the bones preserved (Stan with 65 percent and Sue with 85 percent). The only way to find out how much the new hadrosaur would reveal was to dig. The first few bones collected the year before already indicated that at least part of the tail was possibly articulated, but could it be connected to a whole dinosaur?

At this point in the movies, the paleontologist removes a paint brush and gently brushes the soft, sandy sediments away to reveal a beautiful dinosaur skeleton. If only excavation were as simple as that. Alas, the process of even recognizing a bone in the field can be a difficult task. I have had lengthy conversations with interested folks who have bumped into me on the beaches of the Isle of Wight, asking why I was hanging around on a cold winter's day with an incoming tide. Each time I had to refrain from pushing them off the fossil bones they were standing upon, still partially buried in the exposed beach beds. They simply could not see the shapes that meant so much to me. Actually, I am grateful for their blissful ignorance. Many a beautiful fossil has been destroyed by the inexperienced hand of a person who stumbles over a fossil bone for the first time.

As the team excavated tons of rock from around the site, the outline of a hadrosaur became clearer. As the days turned into weeks, the site expanded. The team carefully picked away around the exposed tail vertebrae. What had started out as the tip of a tale began to look a lot more promising.

As the sediment covering the hadrosaur bones was carefully removed, a strange wrap of iron minerals surrounding the fossil became visible. Tyler had seen similar mineralization before, but not on such a scale. For instance, a skeleton of a small ornithischian dinosaur, *Thescelosaurus*, excavated on a previous MRF field season, had this mineralogy, which was associated with the rarest type of fossil from the Hell Creek—skin impressions. The patience applied to excavating the site was slowly paying off, as was Tyler's cautious approach to the dig process. As the outline of the hadrosaur became more apparent as additional rock was excavated, Tyler realized he had bagged a partial or possibly complete dinosaur. The strange iron mineral covering the fossil made him slow the pace of the excavation even more. On closer examination of the freshly exposed iron minerals, Tyler discovered fossil dinosaur skin impressions—still surrounding the bones. Dinosaur skin impressions, when displayed in museums or depicted in books, often have textures that could not be mistaken by even a casual observer. However, when they are still locked in a tough sedimentary matrix, it is very hard to spot the telltale signs of fossil skin impressions. Skin impressions are often wafer thin, and if not exposed on a fortuitous split along a large piece, they will go unnoticed by untrained eyes. The distinct halo of iron minerals spread across large areas of the hadrosaur.

The block containing the tail of the hadrosaur began to get in the way of the main excavation, and Tyler and his MRF team made the joint decision to remove the four-ton tail block to improve site access. As with any excavation, the combined experience of a dig team often helps when making tough decisions. Ideally, the tail would have been left in place, but that would have made the remaining excavation almost impossible. A natural break near the base of the tail would allow the tail to be parted from what looked to be the dinosaur body block. First the rock containing the tail was covered in aluminum foil to protect the surface. Long strips of burlap were cut, and large bags of dental plaster and quantities of water were transported to the site. The team spent a day wrapping the hadrosaur tail in a protective field jacket of plaster-soaked burlap. In the same way that a doctor uses a plaster cast to hold a broken arm or leg in a static position to allow the bones to heal, Tyler wanted to prevent the fossil bones and skin impressions from moving and breaking. The plaster field jacket binds the rock and fossil into a strong cocoon, allowing large sections of a dinosaur, often weighing many tons, to be safely moved. While the plastering sounds like fun, it hardens quickly, and the pace of work can be frantic. Even worse, as the plaster hardens, it heats up, not much fun when large globs of the stuff are stuck to your skin.

The huge tail block was then parted from the main body block. Where the tail had joined the body, Tyler again spotted the telltale iron minerals surrounding the bones. The deposition of such a mineral was almost certainly a function of the environment that wrapped the dinosaur in a blanket of sands and muds with a "just so" composition. As we will explore later, this mineral deposit was an important key in unlocking the unique circumstances in which the dinosaur was fossilized. Tyler traced the thin brown band around the tail—and it looked to be intact! Was it possible that the mineral associated with skin impressions continued all the way around the dinosaur? The presence of the mineral over large areas of a fossil was unusual, and it might be a marker for more skin, protected by its mineral hard hat.

The plaster field jacket was completed, safely encasing the fossil tail for transport. However, given that it weighed four tons, transport would have to wait. At least the fossil was safely cocooned from the elements.

Excavation on the rest of the dinosaur was halted while Tyler and his team reappraised how they were to excavate this unique find. Could the entire animal be preserved with a complete skin envelope? The problem with any dinosaur find in the field is that it looks like a pile of discolored rock—and that's the good finds! It takes years of experience to know when to stop digging toward a fossil on an excavation. If you don't remove enough of the surrounding rock, you're creating more work for yourself, because you will have to lift more rock out of the site. Ideally, you want to get as close as possible to the specimen without damaging it, reducing the size and weight of the field jacket. Nobody likes lugging back three tons of dirt to find a solitary bone in the middle of an enormous block.

Yet the initial excavation around the tail, coupled with the cross-section of the tail exposed when it was parted from the body block, indicated this was like no other previous dinosaur mummy. The possible layer of skin around the base of the tail appeared unbroken, not collapsed and contorted as is usually the case with the few rare dinosaur mummies known to science. The skin envelope looked like it had been somehow preserved in its three-dimensional form, still surrounding the bones. This did not make any sense, as there was clearly sediment surrounding the bones. It was as if the animal had been opened and sandy mud stuffed inside the skin envelope. There was a paleontological puzzle to solve, and it certainly did not involve some strange prehistoric taxidermist! The environment, dinosaur body chemistry, groundwater conditions, climate, sedimentology, etc. had all combined to create this strange preservational "event." Such a unique find would require a multidisciplinary team of scientists to unlock the secrets of how the dinosaur came to be preserved in such a unique way.

Having seen more of the tail vertebrae, Tyler was pretty sure that the type of hadrosaur he had was *Edmontosaurus*. As described earlier, there were several species of *Edmontosaurus*, so until the animal was excavated and prepared, it was difficult to absolutely identify the dinosaur.

At the end of the 2004 field season, excavation of the hadrosaur site had slowed to a very deliberate pace. The winter months could potentially weather the delicate remains of the exposed fossil, so Tyler made the difficult decision to collect the smaller exposed elements of the fossil separated from the main body block—the tip of the tail, a possible leg, and an arm. Each element was carefully mapped so that their position relative to the body and tail could be reconstructed at a later date. The context of any fossil find is key to reconstructing the what, when, where, and why of things that happened in the past. So much information is lost by hammering a fossil out of its tomb without mapping and logging the site and its sedimentology.

Each part of the dinosaur was wrapped carefully in aluminum foil, then given a protective jacket of plaster-soaked burlap. When the plaster is dry, the burlap gives strength to the field jacket in a primitive but effective way. The main body block was also given a protective jacket of foil and plaster to help it survive the harsh winter. At this stage it was impossible to see what was hidden inside the recovered blocks. Once again, only mechanical preparation of the blocks would answer that question. This was no ordinary dig, where the point of jacketing was defined by exposed bone. The team had to take abstract shapes and mineralogy as locators for when to stop digging. Subtle lateral variations meant that the mineralogy of the sediment changed as they moved across the animal, making it harder and harder to locate the skin envelope. Before long, Tyler and his team were on their knees, faces pressed to the rock, straining to locate signs of the skin envelope. In this situation, it is better to err on the side of caution and leave additional preparation of a surface to the controlled environment of the paleo-laboratory.

The mechanical preparation of dinosaur remains is a very slow, extremely skilled and painstaking process. The dinosaur fossils viewed in museums are often the product of thousands of hours of preparation. On-site, the limbs (leg and arm) and tip of the tail were divided between Tyler and two MRF staff members, Stephen Begin and Tom Tucker. Each piece began the long process of gently removing the encasing rock to reveal the body part entombed within. You would be excused from thinking that Dakota was visiting the dentist when you first saw the tools used by paleontological preparators. A selection of dental drills, picks, resins, and dental plaster are commonplace in a "paleo lab."

Dental drills and micro-jacks (hand-held drills that vibrate up and down rather than rotate) driven by compressed air slowly revealed the enclosed fossil. However, the tough iron minerals that had helped preserve the fossil were making it hard to prepare. The only way forward on some parts of the fossil was to gently tease single grains of sand away from the delicate surface with a pin. Tyler was engaged in this activity when my first e-mails appeared on his computer screen.

I was well familiar with the terrain, having visited it before in my career. I made my first sortie into the Hell Creek Formation with Pete Larson of the Black Hills Institute of Geological

Research. Pete was excavating a *T. rex* (destined for the Houston Museum of Natural Science) from the edge of an ancient pond, now locked in sediment. The Hell Creek Badlands reminded me strongly of the dinosaur-bearing rocks I had worked on in Patagonia, in Argentina. I was also astounded by the numbers of fossil bones lying on the surface, gradually being weathered from their rocky tombs after millions of years. Within a scant few minutes we had collected the fossil bones of turtles, crocodile, mammals, and dinosaurs. An incredible amount of material was being weathered from the steep buttes, rolling down into washouts, dried stream and river beds. The amount of material destroyed by natural processes must run into millions of tons! Some have suggested that the bones of dinosaurs do not weather and erode on the Badlands, but I suggest that they take a hike into the Hell Creek and look for themselves at the unambiguous evidence lying in fragments before them. Of course, if bones are exposed and not excavated, they soon become casualties to the endless process of earth's natural recycling system.

At the time, we were visiting with ranchers Don and Alison Wyrick at their home in Montana, and while there I was able to hike to the topmost part of the Hell Creek and place my finger on the K/T boundary for the first time. A series of coal horizons, followed by a thin clay band, is all that marks this devastating moment in life history. While walking up through the succession of mudstones and sandstones toward the K/T boundary, I spotted occasional fossils poking up out of the butte, but once over the boundary, there was nothing to be seen. The *T. rex* site that Pete and Don had been digging was still in view, so their *T. rex* might well have been one of the last of its kind in the twilight of the Age of the Dinosaurs.

Now my attention had been returned to this promising ground. Rather than mourn the loss brought about by extinction, however, I was filled with excitement by what Tyler Lyson had to tell me. His discovery promised to be very special, in both the world of paleontology and beyond.

In the fall of 2004, Tyler was taking a vacation to the United Kingdom with his mother, and it created an opportunity for us to meet and talk dinosaur face to face. I had only seen a few images of the fossil via e-mail. Now I had the chance to sift through hundreds of images while Tyler explained the details. On the only day that we both had free, I was attending a meeting at the Royal Society on public engagement in science, an area still close to my heart. The Royal Society buildings on Carlton House Terrace are situated centrally in London, making them a perfect spot for Tyler and me to meet, a mere stone's throw from Nelson's Column via Pall Mall.

I waited in the foyer, having spent a morning listening to lectures on how science might better engage with the media, and vice versa. I was keen to talk with Tyler, but I suddenly realized that I had no idea what Tyler looked like. Worse, I was wearing my only suit—albeit with a dinosaur tie. Tyler would probably think he'd found an armchair paleontologist.

I saw a tall figure, with a rucksack and the signature paleontologist field hat, walking toward the Society buildings down Water Place. As the figure drew nearer, I knew he had to be Tyler. Dozens of people had passed in the last half hour, and none looked like they belonged in the field. As he entered the building, I instinctively walked over to greet him. As Tyler and I shook hands, he asked how I knew it was him, and I said it was obvious. "You look like a field paleontologist!" He smiled, relaxed, and sat down in one of the annexes to the reception area. The first thing that struck me about Tyler was how confident and bright he was. This young man in his early 20s had clear direction where he was going in life. Working in academia, you occasionally meet exceptionally talented people, and I certainly count Tyler Lyson among them.

From his rucksack Tyler dragged out a well-used Mac laptop. Since our first e-mail exchanges, I knew that a lot rested upon what I was about to see. We went straight into the image folder to look at the photographs taken during that summer's initial excavation of the dinosaur. Tyler led me on a pictorial guided tour of his hometown of Marmarth, his family, the MRF dig crews, and some of their digs that summer. Before long I was concentrating on the countless images of the hadrosaur dig site. One image stood out, an abstract 3-D form that was unmistakably a dinosaur's body locked in stone. It was as if a sedimentary blanket had been gently laid over the twisted form of an animal: a smudged outline of an arm protruded alongside a leg and a cone-like structure pushed back from the main block, looking like it could only be a tail. Astoundingly, I was not identifying the limbs and tail from the skeletal anatomy. These were the shapes of the limbs. I looked at Tyler in disbelief, and he only smiled, insisting, "I told you this was a special fossil."

Tyler and I talked for hours, until we had outstayed our encampment in the Royal Society. We decided to head for dinner at a pub just off Trafalgar Square to continue talking about dinosaurs. I was impressed by what a focused, mature, and obviously intelligent person Tyler was, and he has subsequently continued to impress. We parted company after dinner with a clear plan in mind. We agreed that by joining forces, we could raise suitable funding for the excavation and preparation of the fossil. The game was afoot.

Over the next few weeks and months Tyler and I began the process of planning a reconnaissance trip to North Dakota so that I could see the fossil firsthand. Tyler also wrote about many of the additional exciting finds from his family's land, but the one that continued to stand head and shoulders above the rest was the dinosaur with skin impressions.

Finally, a short space appeared in both our hectic schedules where we could meet in the North Dakota Badlands. On a cool and damp morning in October 2005 I set off from Manchester, wondering if Tyler had found a fossil that might change the way we view dinosaurs. When he had explained a three-dimensional skin envelope surrounded the tail, all I could think of was an image of a dinosaur steak! While I knew the soft tissue was long gone, a skin envelope that had the potential to constrain so many soft-tissue parameters continually raced through my mind. I flew from Manchester to Chicago, then Chicago to Denver, and finally in a jumpy prop-plane from Denver to Rapid City, South Dakota. I still had a 200+ mile, four-hour drive from Rapid City, across the seemingly endless prairie from South to North Dakota.

As you drive across the prairie, buttes rise like the backs of sleeping giants in a grassland sea. Many of these buttes and underlying rocks are rich in dinosaurs, so I can be forgiven for thinking of giants while heading north on a road that is so straight, you can recalibrate your compass with it. The first snow had already fallen the previous week, but subsequent bright sunshine had melted most of the early snow, making an easy journey north. Just when I could not imagine another section of straight road, the lights of Bowman, North Dakota, appeared on the horizon.

The road west from Bowman to Marmarth gently dips and rolls over the prairie, passing occasional exposed patches of Hell Creek Formation. Just before I reached the outskirts of Marmarth the road dropped down into the Badlands, leaving the prairie behind. A life-size welded-steel *T. rex* stands on a hill just east of the town. It seems quite fitting that visitors are greeted by a dinosaur. It was designed and built by Tyler's brother, Derek, and a colleague, David Shepherd. I imagined dinosaurs were a big thing in the Lyson family. After crossing the Little Missouri River, I arrived in Marmarth, North Dakota.

Tyler's hometown used to be a bustling place with a population of more than 2,000, but it now hovers around the 140 mark. The town was established by the railroad in 1907 and named after Margaret Martha Fitch, granddaughter of Albert J. Earling, president of the Chicago, Milwaukee and St. Paul Railroad. A combination of a flu epidemic and the railroad no longer stopping at the town had a devastating effect on the community's prosperity. However, the community spirit that remains is both warm and welcoming. No sooner had I stopped in Marmarth than a friendly passer-by pointed me toward Tyler's House. I liked this town a lot already.

It was great to see Tyler in his home environs. What a perfect place for a dinosaur hunter to have his roots: smack-dab in the middle of one of the richest dinosaur hunting grounds in the world. The surrounding buttes and valleys around Marmarth were also the perfect age, 67-65 million years old. I was greeted warmly by Tyler's mother and father, whom I would soon discover made the best tea and margaritas, respectively, in North Dakota. That night I slept very soundly.

The following day, Tyler and I had a typically early start. Paleontologists have a habit of using every scrap of the daylight hours available to them. We jumped into a 4x4 truck and headed west out of Marmarth. Soon we left the tarmac road, turning north on a dirt track, which soon disappeared. This wasn't the first time I had fully understood the bone-shaking reality of the term "off-road," having spent many months in the field from North to South America, Europe and North Africa. I am still surprised that car rental companies have not yet added a tick box for, "I declare that I am a paleontologist and will not attempt to drive this vehicle to destruction while climbing vertical rock faces in the pursuit of dinosaurs."

As we drove toward the site, for miles around all you could see were beautiful, dinosaur-rich Badlands. Melting snowdrifts dotted the landscape, but fortunately the ground was still dry enough for our vehicle to make headway. A simple and very important equation to remember, I was to learn, is: Badlands + water = trouble, whether that be from rain or snow. The mud that makes up so much of the Hell Creek Formation becomes impassable—hence they are bad-lands to traverse, farm, or live on.

After a long bumpy journey to the middle of nowhere, we arrived at a group of small buttes that looked like so many we had already driven past. When Tyler turned the pickup's engine off, we were greeted by complete, almost deafening silence. The scent of sagebrush was heavy in the air. We walked with deliberately heavy steps, making as much noise as possible. That way you're less likely to come across a snake. The site was still a hike from where we had stopped. We scrambled up and down weathered, barren slopes. I was trying hard not to walk ahead of Tyler, but this was about as exciting as dinosaur hunting gets. I had never seen dinosaur skin impressions in the field.

As we rounded a small butte I spotted another snowdrift, but not shaped like any of the melting ones we had passed that morning. As we drew closer, the pile of snow began to look more like a giant pile of plaster of Paris. It was the winter field jacket that Tyler and his team had carefully wrapped around their precious find at the end of the 2004 field season. The outline of a dinosaur was not obvious, but I was still completely lost for words. We both stood there for a minute, grinning knowingly at each other.

Tyler explained that the fossil might well preserve the three-dimensional skin impression of the dinosaur body, unlike so many dinosaur fossils that had collapsed into a jumble of skin impressions and bone. Viewing the block that contained the carcass, I found myself wondering if the body shape could be preserved intact. There was a chance that Tyler and his team had discovered one of the rarities of the fossil record, a dinosaur mummy.

The four-ton tail block that Tyler and his MRF team had already removed was parked a short distance from the main body block. Tyler walked me around the cone-shaped block, explaining how the bones and skin were distributed within the field jacket. As he explained how the cross-section at the base of the tail showed a skin envelope around the bone, I did a double-take. If the skin had not collapsed against the bones, how could such a unique fossil be formed? The Sternberg mummy had skin impressions pushed up against the bones, often forming discrete ridges and flaps of fossil skin, but nothing as complete as Tyler was indicating for Dakota's tail regon. Leonardo had a completely different type of skin preservation, with almost a trace "outline" of the scales that had once been, now delineated by lacelike ridges of sandstone. Dakota possessed very much a distinct structure with volume. This was no skin or spiderweb-like trace: Dakota's skin also had depth, only a few millimeters at best, but this looked like surface-relief, mineralized skin, not an impression. The beautiful hand of the Senckenberg hadrosaur mummy came to mind, with more surface relief to the skin, but that was still pushed up against the bones of the hand.

Tyler and I began to plan what we would need for next summer's field season. The division of workload between Tyler and me was clear. Over the next few months I had to build a team with the right skills set to investigate the hadrosaur fossil, while Tyler assembled a field crew for the excavation and extraction. We each had plenty of field experience, but Tyler would manage and organize the excavation, while I sorted sample collection, mapping, and logging in and around the dig site. As we paced out the dimensions of the body block, ten feet by six feet at the widest point by a maximum of five feet thick, we knew it would take a huge field jacket to recover the body block (with, we hoped, a neck and head inside) in one piece. The skull of any dinosaur helps in confidently diagnosing the species, as the post-cranial skeleton of many dinosaurs, especially hadrosaurs, can be hard to differentiate.

Tyler produced a small bundle of clothes from his pack. As he carefully unwrapped it, he smiled, knowing what my reaction would be. In the palm of his hand, Tyler carefully held a golf-ball-sized piece of sandstone. Its surface was covered in regular scales, not the mere trace fossil of skin impressions, but fossilized skin with real depth and structure. As the sunlight struck the surface of the fossil, minuscule canyons meandered between the raised scale platforms. It was as if the scaly back of an animal had been dissected and turned to stone, retaining all its perfect detail. The best I could manage was "That's incredible!"

What kind of edmontosaur had Tyler found? It was probably *Edmontosaurus annectens*, but until the fossil was excavated, prepared, or CT scanned, we would have to wait for an answer. Patience is a necessity in paleontology, but I was ready to start that very morning.

CHAPTER SIX
PREPARING THE CAMPAIGN

"She sells seashells by the seashore.
The shells she sells are surely seashells.
So if she sells shells on the seashore,
I'm sure she sells seashore shells!"

MARY ANNING IS IMMORTALIZED in the playful tongue twister that tells of her fossil-hunting exploits at the beginning of the 19th century. Academics, artists, philosophers, and enthusiasts have collected fossils for many centuries. The amateur collector and dealer have played an integral role in finding and making fossils available for study and contemplation. The origin of the science of geology at the turn of the 19th century was fueled by partnerships between collectors, dealers, museums, and academics. The process of finding, excavating, and identifying the organisms locked in stone has changed little since Anning's day, though the invention of power tools and excavators has greatly aided the modern excavation process.

The acquisition of paleontological specimens from the field is also an important role for museums and universities, but the time and resources required to find, extract, conserve, and store specimens are often not available. The time needed to find specimens is certainly one of the key areas to address, when discussing the acquisition of paleontological specimens, as the time required to locate new specimens in the field is disproportional to the time available to most museum and university-based paleontologists. The work commitment of curators, lecturers, and researchers rarely allows the indulgence of prospective fieldwork. The finds attributable to such workers are often restricted to "lucky" finds or are a response to other collectors or members of the public who have located material in the field. The reaction of museums and universities to such finds is then dictated by the budgets and time available to excavate, prepare, conserve, and store such finds.

When a rare fossil is offered for sale, members of the scientific community, who have a research interest in the specimen, often raise the issue of "ethics." The ethics of a sale should be restricted to establishing and clarifying a specimen's provenance, authenticity, whether permission to excavate was granted, and the legality of ownership. Any public institution or private individual should rigidly apply these criteria to all prospective acquisitions, irrespective of the importance of the specimen. Only when all of the acquisition criteria can be proven should a specimen be offered for sale or purchase.

The excavations of large fossil specimens, usually vertebrates, are sometimes beyond the financial resources of private collectors. In such cases a museum or university can and should work in partnership with collectors (amateur, professional, or commercial) to retrieve such specimens. The field experience and expertise of collectors should be viewed as a resource and not as threat to our paleontological heritage.

The high quality of preparation work attained by many private collectors is also an untapped resource for many museums. An example of such preparation, displaying an intimate understanding of the fossils prepared, is the work of Mike Marshall from Sandsend, near Whitby. He has collected fossils from the Yorkshire coast for much of his life, which has led to an appreciation of the material he prepares, delivering unparalleled results rarely observed in fossil preparation. If you ever see a beautiful ammonite from the Lower Jurassic of Yorkshire, usually a *Hildoceras* or *Dactylioceras*, think of Marshall, as the chances are, he was the chap who lifted the beast out of its grave.

The enthusiasm that collectors exude when talking about specimens, localities, or collections is something that some people reading this book might have experienced. Such enthusiasm is infectious. This passion displayed by so many collectors makes them perfect ambassadors for fossils and paleontology, and they are often the primary point of contact for young fossil hunters who have yet to make their first great find. The enthusiasm of many collectors informs newcomers that all finds are of great importance. This is a significant lesson to learn before any formal educational process can begin to breathe life into these ancient remains.

Local collectors have aided almost all of the many excavations I have undertaken on the Yorkshire coast and Isle of Wight. One collector, whom I probably owe my life to, is Keith Simmonds from the Isle of Wight. Many years ago I was happily hacking away at an overhanging boulder on a rising tide on the island's southwest coast, oblivious to the instability of the large slab of rock immediately above my head. Simmonds, who was taking his daily stroll along his favorite cliff section, dragged me away from the *Iguanodon* vertebrae that I had completely focused upon, and suggested that I revise my route of attack. Collectors like Simmonds have brought us many exciting dinosaurs, such as *Neovenator salerii*, and he continues to devote thousands of unpaid hours to the hunting, excavation, and preparation of fossils. I am sure he will also continue to help greenhorn academics along the shores of the Isle of Wight for many years to come.

The relevance of collectors—Victorian to present-day—is key to understanding the skills and planning required when running a successful dig. The planning of a dig takes experience, patience, and money. All potential bases need to be covered before you even get into the field. Once you are in the middle of nowhere and a particular tool or material has not been purchased or packed, it can hold up the entire process.

The first six months of 2006 were hectic for both Tyler and me. He was studying for his finals at Swarthmore College, and I was teaching a new vertebrate paleontology course at the University of Manchester. When we had time, we exchanged e-mail messages and phone calls, continuing the planning for the summer. We had to make sure that during the excavation we recovered every shred of evidence that might help reconstruct the life, death, and fossilization of the dinosaur mummy. In recent studies organic molecules had been extracted from the fossil bones of *T. rex* and the woolly mammoth (specifically collagen), so we had to avoid contaminating potential sample sites on the mummy. For instance, various glues and compounds are often used to consolidate fossils during excavation, mostly to hold together crumbling bones when first exposed after millions of years. Yet the use of such material destroys any chance of recovering uncontaminated organic molecules from the fossil, since most glues contain organic compounds. Even handling samples would leave traces of our own tissues (e.g., skin cells, hair) on the fossil, again impacting any organic molecules recovered.

We would have to excavate the fossil in such a way that as much of the skin envelope as possible remained "sealed" in its sedimentary cocoon. Samples that were to be used for the potential recovery of organic molecules would have to be extracted in a clean-room, laboratory-controlled environment. Keeping as much of the skin envelope covered in matrix would also help prevent damage to the delicate surface textures that had already been exposed in a few small patches in the earlier field season. When the tail block had been parted from the main body, we had observed that the skin envelope was less than 0.5-millimeters thick in many places. This was too delicate a structure to expose to the brutal excavation environment of the Badlands.

Retaining a sediment coat around fossil remains has both advantages and disadvantages. The chemistry of the sediment today has the potential to yield information on the pore-water chemistry that existed more than 65 million years ago around the carcass. By retaining as much of the sediment around the fossil as possible, potential information that might otherwise have been chopped away in the field can be analyzed later in the lab. The communities of bacteria that thrive on the decaying remains of organisms would themselves have produced waste products that affected the preservational environment and the sediments in which they lived. The species of bacteria present would in turn controlled by the prevailing environmental conditions. Different type of bacteria favor oxygen, methane, or anoxic conditions. It is sometimes convenient to talk of sediment "packages," separating such subtle changes by separating the sedimentary environments (facies) from one another. The change in sedimentology can range from over-bank muds on the side of a river system to the crevasse splay (sheet sands) of a flooded plain spewed from overflowing rivers to the multi-tiered channel sands with their intricate sedimentary structures. They all reveal much of the past world now locked in stone. The sedimentary samples collected in the field and from those still encasing the mummy would help reveal these complex conditions and how they might have changed over various parts of the body as the dinosaur decomposed. The 3-D digital outcrop maps we intended to generate using a laser scanner, when combined with sedimentary samples and field observations, would provide a more accurate window into the world of our dinosaur.

We also had to make sure that the excavation process kept both the dinosaur and the dig-team members in one piece, not necessarily in that order. Health and safety have even made their way into the depths of the Badlands. In the excavation of any fossil, a large number of power tools, pickaxes, knives, scalpels, and other sharp objects are used, not to mention winches, chains, and front-end loaders. Having one person in control is essential, and Tyler would assume that role. This is not to say he would not take advice from his field crew, with their wealth of experience. He was selecting many seasoned field hands to work on the mummy site.

The division of labor between Tyler and me into excavation and science teams seemed a logical step to take. We were working closely on both fronts, so in many respects it made little difference to the team dynamics; it just made organization so much easier. While Tyler managed the MRF volunteers on the dig site, I would choreograph the arrival and departure of the many scientists working on the project. The University of Manchester team would include paleontologists, biologists, geomicrobiologists, sedimentologists, stratigraphers, digital mappers, material scientists, CT scanner experts, computer scientists, and many more.

The research team included Rob Gawthorpe, Dave Hodgetts, James "Joe" Macquaker, Kevin Taylor (Manchester Metropolitan University), Jim Marshall (University of Liverpool), and Franklin Rarity working on the sedimentology, isotope geochemistry, and digital mapping (using a laser scanner) of the site. All of the team except Marshall would work at the dig site that summer. Luckily, I had worked with the entire team before and knew them well. All told, the team had a broad wealth of experience and would make a rigorous interpretation of the site and its sedimentary environments. The myriad of disciplines to be applied elsewhere in the project would be concentrated on the preparation and lab stages.

One of the key techniques that I wanted to apply to the mummy excavation was the use of LiDAR (LIght Detection And Range) to generate 3-D digital outcrop maps of the dig site and surrounding landscape. We had experimented with the technique on paleontological sites in the Spanish Pyrenees the summer before. At Fumanya, a site north of Barcelona, my team had worked in collaboration with a Spanish research group to produce a huge 3-D map of thousands of dinosaur tracks striding across the uplifted, near-vertical walls of a mountain. For the first time a paleontological site had been accurately mapped, providing a valuable record of the relative position, condition, and geometry of the tracks.

What is LiDAR exactly? Light detection and range imaging is a highly accurate method of acquiring 3-D spatial data that has been widely applied in other areas of heritage conservation, especially archaeology. LiDAR has been used in many related fields (e.g., engineering, surveying, atmospheric physics) for several years now. To date, though, LiDAR has been underutilized in paleontology, both as an analytical and conservation tool. Brent Breithaupt has used terrestrial LiDAR imaging and digital photogrammetry separately to record and map small sections of outcrop in Wyoming and Colorado containing abundant dinosaur tracks and skeletal remains. The potential to integrate LiDAR and photographic data and collect high-resolution quantitative data from sites through remote surveying suggests the method may provide a means to merge conservation with the scientific exploration of heritage sites.

The fully portable RIEGL LMS-Z420i 3-D laser scanner was chosen for its ability to rapidly acquire spatial data under demanding environmental conditions. Another key factor in the choice was that my department had recently acquired such a machine. The LiDAR has a range of 800 meters, 80° vertical and 360° horizontal fields of view, and can even be powered in the middle of nowhere by a 24-V or 12-V car battery. The scanner uses a near-infrared laser that is safe on the eyes and requires no additional safety precautions. A ruggedized notebook computer, combined with a suitable software package (RiSCAN PRO) enables an operator to acquire, view, and process 3-D data in the field, thereby increasing the level of quality control on survey data. A digital camera (6.1-megapixel Nikon D100) was mounted on the scanner and calibrated to provide images that produced a photo-realistic representation of the excavation. Precise global positioning was provided by a Trimble Differential Global Positioning System (DGPS).

The LiDAR works by firing a near-infrared laser at the surface it is surveying. The distance from the rock surface is automatically calculated by feedback from the laser; this process generates 12,000 data points every second. The result is a "cloud" of X, Y, and Z data points that represent the 3-D surface of the rocks being scanned. Quite stunning. Because the LiDAR is linked into a GPS (global positioning system), the team can move the LiDAR from one scan station to the next, scanning large areas, but still be able to "stitch" together the final 3-D surface maps that they generate using the GPS data. Any interpretation of the sediments we were studying in and around the mummy could thus be seen in the wider context of the 3-D outcrop model. This was a powerful tool for our investigation.

The scanner could be fitted with either a vertical mount or tilt mount. We used the tilt mount in the Fumanya survey and would also use it for the mummy site; it provides a full 180-degree rotation from the horizontal, giving a very wide field of view. The tilt-mounted scanner, digital camera, and GPS were mounted on a heavy-duty surveyor's tripod, making it a cumbersome handful to lug around on mountainsides and Badlands.

Full coverage of the mummy site required a number of scan stations dotted around the excavation. Multiple scan stations provide more detailed 3-D shape information by eliminating shadows (i.e., areas not visible to the laser) in the data caused by irregularities in the surface. Both perpendicular and oblique scan perspectives were therefore necessary in this instance to prevent shadows occurring within the landscape, which form features of negative relief (i.e., trench, deep valley, etc.) on the scanned landscape.

This array effectively merges scan information with undistorted digital photographs to produce a high-resolution, 3-D photo-textured model. Before photo-texturing, scans must first be triangulated—that is, a surface must be created by connecting adjacent points with triangles. Once triangulated, the scans consist of a series of points that have been connected to form a triangulated mesh. Each individual pixel within the photographic image is linked to its X, Y, and Z coordinate within the correct triangular vertex by a texture coordinate recorded within the triangulated mesh.

The accuracy and resolution of 3-D surface geometry represented in the textured surface largely depends on the number of points within the scan data (i.e., the scan resolution).

Photo-textured models are more likely to accurately depict 3-D surface geometry when derived from dense high-resolution point clouds. It is also desirable to use merged scans of the same exposure recorded from different perspectives (i.e., different scan stations) to reduce pixel-stretch effects. Irregularities in the scanned surface mean that point clouds recorded from a single-perspective are inevitably 2½-D in nature and contain areas not represented by laser points (i.e., shadows). Integrating scans from different perspectives helps to fill previously vacant 3-D space, effectively eliminating shadows and the need for pixels to stretch to match adjacent scan points.

Our use of LiDAR at the mummy site would be the first time the Hell Creek Formation had received such surveying accuracy. The weathered buttes of the Badlands would test the technique to its limit, but would ideally provide some unique contextual data for the mummy team. The team would have to systematically recover every fragment of evidence to answer the tough questions locked in the fossil for more than 65 million years. No stone or even grain of sand would be left unturned.

The mummy team also needed funding. The cost of such an expedition mounts very quickly. I began with the University of Manchester and then the National Geographic Society (both the Committee for Research and Exploration and the Expeditions Council) in Washington, D.C. When I first contacted National Geographic, everyone who heard the story of the dinosaur was very excited. The exposed skin and section through the tail already indicated parts of the hadrosaur possessed an uncollapsed skin envelope. They agreed that this offered vast research potential. Before long, I was busy filling out grant proposals to secure funding for the science project from National Geographic. Fortunately, I could draw on a wealth of experience from the team, which made relatively short work of explaining our proposed excavation and research program.

By June 2006, I was ready to make my second reconnaissance of the hadrosaur site in North Dakota. Tyler was home in

Marmarth from college for a few days, and I had some public talks to deliver in South Dakota. We could both free up some time to see how the mummy site had survived a second winter.

This time when I landed in Rapid City, it was warm and raining. I had brought some Manchester weather with me. June is one of the few months in the Great Plains when it rains, heavily, and this was a worry. Would Tyler and I be able to tour the site? As I drove from South to North Dakota, however, the clouds began to clear. In my rearview mirror I could still see the dark clouds over the Black Hills, but to the north were clear skies. Leaving the Black Hills, I passed from the realm of the whitetail deer (*Odocoileus virginianus*) to that of the pronghorn (*Antilocapra americana*). Compared to pronghorn, deer are not smart, as empirical evidence showed. The former are quite excellent at avoiding cars (although roadkill suggests one or two slip through the filter), but deer seem to be attracted to cars like bees to honey. This makes driving in the Hills hazardous, especially at dusk and dawn.

After another back-numbing drive across the prairie, the lights of Bowman were a welcome sight. Before long I was on the Marmarth road, which gently sank into the Badlands. The familiar *T. rex* sculpture that Tyler's brother had built greeted me as I reached the outskirts of his hometown.

The excavation we were planning for July was to be the final push to recover the entire hadrosaur. The team would have to remove the enormous ten-ton block containing the body and, with luck, the skull. The painstaking task of collecting countless rock samples also had to be coordinated, as did the LiDAR mapping of the site.

As Tyler and I sat down to plan the dig, the first thing we realized was that we would be adding at least 30 people to the residents of the town—a 20 percent increase in the population! The headquarters for the excavation and science team would be the old Milwaukee Railroad Bunkhouse. It still functions as a rooming house with dormitory-style rooms, attracting bird watchers in the spring, who come to watch the sage hens dance.

In summer it hosts the influx of paleontologists and dinosaur enthusiasts. We needed to book every room in the Bunkhouse that July.

Tyler and I went over the teams he was assembling for the dig and that I was bringing over to map the site and collect sediment samples. Everything was planned to the last sample bag and sack of dental plaster. Finally, we drove out to see the hadrosaur fossil, the focal point of such intense labor. We both had been too busy to return to the site since our snowy visit in October of the previous year. The rain that had been falling in South Dakota had not reached as far north as Marmarth, so it was safe to make our journey across the Badlands to the mummy site.

We jumped into Tyler's pickup and headed west out of town, once again crossing briefly into Montana before heading north-northeast and back over the border into North Dakota to the site. The Badlands looked different from when I had seen them last October. The amber glow of autumn had been replaced with the sharp light of June. The change in season discombobulated my sense of direction, because the Badlands looked so different. But then the buttes surrounding us began to look a little more familiar, and Tyler announced that we had arrived. We soon stood examining the body block. It had survived the harsh winter.

We worked our way around the site, sorting sampling areas, scan stations for the digital-mapping equipment, and site access for the front-end loader that would have to carry the huge body block off-site. After a few hours of concentrating on mummy matters, we were all set. The start date of the excavation was July 5, 2006—Dino Day!

CHAPTER SEVEN
EXCAVATING A PAST WORLD

"The poetry of the Earth is never dead."
—John Keats

WILLIAM "STRATA" SMITH, in the first half of the 19th century, used fossils to trace the surface collage of disjointed parcels of rock from one end of the British Isles to the other. His work successfully provided the underpinning principles for the science of stratigraphy (specifically biostratigraphy), showing that short-lived fossil species could provide clear markers for specific ages of rocks. These fossils are now known as zone fossils (or index fossils). Some species were restricted to very short periods of geological time, making them ideal zone fossils, especially if they were abundant (ideally globally). With Smith's guidelines, geologists could, for the first time, use information from the fossil record to locate where they were in geological time. That skill was especially useful when hunting for Carboniferous-aged coal to fuel an industrial revolution.

The occurrence of new species in the fossil record provides useful stratigraphic markers that can be employed to assert a specific age for a "package" of rock or sediment. However, landscapes can provide many obstacles to the interpretation of such packages by their sheer size and complex fusion, since each succeeding deposit of sediment alters the evidence of the prior.

Subsequent weathering and erosion further blurs the sedimentological and stratigraphic picture.

As you scan an intricate landscape, your eye picks up the multitude of complex shapes, shades, flora, and fauna. In the desolate Badlands of North Dakota the underlying rocks paint a colorful three-dimensional image of the Late Cretaceous rocks laid down more than 65 million years ago. How we interpreted these past landscapes would rely heavily upon the next month's fieldwork.

Eighteen months has passed since I had first been in contact with Tyler. Once again I found myself on a long-haul flight from London to Chicago, but at least this time I could take a direct flight from Chicago to Rapid City. When I arrived, an impressive lightning storm was breaking over the Black Hills in the distance. One of my Ph.D. students, Chris Ott, greeted me with a beaming grin from behind a rather large growth of beard. He and I were the first members of the Manchester team to arrive. Chris had already been in the field for a month, scouring the Hell Creek Formation for data for his doctorate. Chris thought it was quite ironic that he was meeting his British supervisor in the field for the first time on July 4, America's Independence Day, since Chris is from Wisconsin.

We stayed in nearby Hill City on July 4 before driving the 200+ miles north to Marmarth. Waking at 5:30 a.m., courtesy of the time difference between the United Kingdom and the United States, I helped pack Chris's "transport." He is very proud of his field vehicle. I, on the other hand, was curious how it was still on the road. The well-used 1978 GMC Suburban with a 454-cubic-centimeter big block engine was a monster, although it had lost some of its teeth. I think the Suburban was once white, but years of fieldwork and hundreds of thousands of miles had taken their toll. Nonetheless, I had to be admit that it was the perfect transport for paleontologists and a huge pile of field equipment. We never had to worry about dirt and scratches, and it offered a payload capacity to shift a reasonably sized dinosaur—this was a great field vehicle.

Once more we made the trek across the prairie from South to North Dakota. Before I knew it, we had crossed the Little Missouri River and arrived in Marmarth. Chris and I headed for the MRF prep lab, as we had arranged to meet Tyler there. The moment the roar of the Suburban engine died, we could hear the angry bee hum of prep tools making their mark on blocks of resistant stone. Walking past the large windows of the prep lab, we spied spectacled and dusty folks intently gazing at lumps of rock they were working on. Tyler was holding a fossil turtle skull and talking to one of the MRF staff, Lou Tremblay. They both looked up and gave spontaneous grins. It was good to be back in Marmarth.

We left Tyler and his team and headed back to the Bunkhouse to unpack the Suburban. As we unloaded our field gear and luggage, members of the MRF field crew began to return from the day's digging. We were soon exchanging stories of dinosaurs from Patagonia, China, and Europe. Paleontology can make the world seem awfully small sometimes. After unpacking the Suburban, talking, and sorting, I sat down to write my field journal; it was 11:22 p.m. mountain time and 6:22 Greenwich mean time . . . bedtime.

The next morning, we got off to a 6 a.m. start. Yes, paleontologists like to start digging early. Breakfast was served promptly at 6:30 in the Marmarth Café. Looking around the tables, it was clear that the MRF field crew were all excited about the coming weeks. It doesn't matter if you're a seasoned field hack or a first-timer to the Badlands. The anticipation and eagerness to get into the field were tangible. Many spoke of a good night's rest. Others were a little stiff from the microscopically thin mattresses in the bunkhouse. Others slept courtesy of a nightcap from the local watering hole, Pastimes. The clatter of cutlery on plates, smell of cooked breakfast, and field banter filled the air of the café. Every one of the crew would have to drink plenty of water, as the day was already warming up fast. Finally, Lou Tremblay stood up with the site-allocation sheet and read off where each digger would be working for the day. He and Tyler had already worked out the digging schedule for the next few weeks. That day only Tyler, Chris Ott, and I would head for the hadrosaur site. The remaining MRF crew were destined to dig on two separate *Triceratops* sites and Tyler's favorite turtle site.

As we drove over the Badlands to the site, I was amazed at how parched the landscape looked. Brush fires were raging to the north and we could soon see the haze of smoke on the horizon. Tyler explained that the area was like a tinderbox, as they had not had a decent rainstorm in many months. As we drove, a huge dust cloud billowed out behind us. Apparently, it had been a good year for rattlesnakes. The relatively mild winter had allowed many more to survive and breed, especially with a plentiful rodent population, which had also taken advantage of the relatively mild winter. The hemotoxic venom they are capable of delivering makes them potential field problems for all crews. However, the distinctive rattle, a series of modified scales at the tip of the tail, usually gives plenty of warning. I was more worried about the very young snakes, which do not have a rattle and cannot gauge how much venom to pump into their prey, often delivering too much—small snake does not equal less venom! The availability of anti-venom has reduced the fatality rate of rattler bites to a mere 4 percent. Yet I didn't care to become a statistic.

As we arrived at the site, we walked down into the small valley that nestled between two buttes and spotted the hadrosaur mummy body block. The vigorous staccato rattle of a disgruntled rattlesnake greeted us. We had a resident rattler at the site trying to warm itself at the beginning of its day's hunting. Tyler laughed and said the snake would probably not come back, what with all the activity at the site.

We started working, identifying and preparing areas where we could set up the digital-mapping equipment. We walked, stumbled, and climbed in the surrounding Badlands till we were sure we could cover every possible angle required for the mapping process. The LiDAR required a clean line of sight for each nook and cranny of the landscape, so locating clear vantage points was important prior to the LiDAR team's arrival. Forethought would save valuable mapping time for the team. The sun was still not at full height, but the temperature was already into the 90s. The excavation was going to be hard work in the heat, as there was little to no shade at the site.

Over the next few days Chris, Tyler, and I and a couple of willing volunteers from the MRF helped to dig trenches below, around, and above the hadrosaur mummy. The walls of the trenches were shaved to a clean, flat surface, so we could study cross-sections of the sedimentary structures preserved in the rocks. When sediments are originally deposited, they leave telltale clues as to the type of event in which they were laid down. We also collected samples of the sediment below, within, and above the level that held the fossilized dinosaur. The samples would be fashioned into polished thin sections (a very thin layer of sediment stuck on a glass microscope slide) so that we could study how the chemistry of each sediment layer changed relative to the position of the fossil.

Chris and I soon completed our initial survey and sediment sampling of the site, so he disappeared into the Badlands looking for additional sedimentary clues, while I joined Tyler and his MRF team. We were ready to start undercutting the hadrosaur mummy. During this procedure tunnels are drilled underneath the giant body block to start the process of eventually freeing the dinosaur from its stony tomb. We did not want to move the body block until the digital mapping team arrived from Manchester, but we could get a head start on what would be a long, hot extraction process.

We spent several days drilling rock beneath the huge body block. It was tough, arm-aching work. By the end of each day, we were sore all over. One of the team members was less fortunate than the rest of us. Steve Cohen, an MRF veteran, was lying down on the opposite side of the body block from me as we were feeding steel support rods through the tunnels beneath the hadrosaur. All of a sudden Steve let out a loud "Ouch!" Quickly standing, he reached around to his back, trying to swat something that had crawled up his shirt and stung him. A small scorpion had found its way into Steve's shirt, and when he had moved, it started jabbing its venomous sting into his back. I had Steve lift his shirt: Fifteen stings were clearly visible across his back, very painful. Three of the stings began to swell more than the others, where the scorpion had obviously emptied poison into his back more efficiently.

We had Steve drink plenty of water, and we watched the lumps on his back angrily rise and then thankfully recede in size. Steve was lucky not to suffer an allergic reaction to the sting. If he had, it would have meant a frantic race across the Badlands to Baker, Montana, which was 20 miles away. During those minutes of anxiety the distance seemed a lot farther.

Soon the digital mapping (LiDAR) team arrived from the University of Manchester. They had five days to map a few square kilometers of the Badlands. The team was used to working on outcrops for the oil industry, so persuading them to put a large chunk of their field-time budget to the mummy excavation was a great boon for the project. The team consisted of Rob Gawthorpe, Dave Hodgetts, Franklin Rarity, and Kevin Taylor (a sedimentologist from the Metropolitan University of Manchester).

Once the survey equipment was set up at a locale, Hodgetts would patiently run through the survey protocols in the software that he himself had written. The gentle hum of the LiDAR could be heard as the oscillating mirrors that scattered and collected the light from the emitted laser spun furiously in the drum housing of the apparatus. The upper half of the unit slowly rotated, scanning the visible landscape. At the end of each scan cycle, the unit would rotate once more, clicking images with the camera mounted on top of the unit. This often necessitated a strange dance around the LiDAR by any onlookers, who were all trying hard not to be image-grabbed into the scan. Once the 360° scan cycle had been collected, the whole unit had to be shut down, bagged, and carted to the next scan station. Carrying the LiDAR unit, heavy-duty car batteries, tripod, and support equipment made for strenuous work in the heat of the day. The fact that the LiDAR required a vantage point for each scan meant the crew had no respite from the unrelenting sun.

After five days of moving LiDAR gear up and down Hell Creek buttes, the team had completed their survey of the hadrosaur body block and surrounding landscape. They took the rest of the day to pack the equipment safely for travel. Tyler and the MRF team were able to shift up a few gears in terms of speed and start preparing the body block for extraction, a term that is quite apt given the copious amounts of dental plaster being used.

As the MRF team started brushing off the body block for its plastering, I picked the last few sediment samples from its surface. Each sample was photographed in place before being chipped from the surface. As soon as I finished, aluminum foil was unfurled from numerous containers. The body block soon resembled a giant Thanksgiving offering. Once Tyler was happy with the coverage of foil, the shining mass was bound with strong tape to hold the foil in position. We then all got completely plastered! The main task was to make sure the body block had a decent layer of plaster and burlap. Since the weight of the block would be in excess of ten tons, the field jacket would have to be very strong. Layer after layer was slopped onto the body block. Each layer of plaster and burlap added more weight, but also additional strength, to the block. The temperature at the site reached a searing 126°F, hot enough to fry an egg on the hood of one of the field cars. The heat meant we had to work that much faster, as the plaster would rapidly begin to harden as soon as it had been mixed. Before long we had a production line of plaster mixing and "goop" application. Goop is the thicker plaster mix used to smooth over the burlap strips that wrap the mummy. I have to say, slapping goop onto a field cast and smoothing the rough plaster to an even finish is a very satisfying experience.

During the process I stopped to consider that we were wrapping bandages around a vast natural animal mummy, and wondered what the ancient Egyptians would have made of us. Possibly they would have suggested we had been in the sun a little too long.

After a few days of plastering we had finished the top half of the mummy block. It still had to be plastered on the underside, which included expanding the tunnels we had drilled earlier beneath the dinosaur to take the steel frame that was ultimately going to support the block. As we continued to plaster around the overhangs of the body block, Norman Gardner, another MRF veteran, lay on the site floor, enlarging the tunnels. Poor Norman had to put up with Dan Pepe, a Yale Ph.D. student helping out on the dig, and me slopping plaster over him. We also successfully managed to fuse ourselves to the field jacket, several times. You see, when you are holding plaster-soaked burlap on the underside of the block, you have to let the plaster dry while you support the burlap. This is when gravity plays havoc with your skin. Let's just say that hairy people should avoid plastering, for every time we pulled hands and arms away from the body block, we seemed to leave hair and skin behind, Dan more than most! Tyler, Steve Cohen, and Tom Tucker, another MRF staff member, were busy doing the same procedure on the other side of the block—with a frantic, plaster-splattered Doug Hanks, also MRF staff, keeping up with the plaster and goop supplies.

The LiDAR team were stowing away the last few pieces of equipment when a lone figure appeared, like a pale rider on the horizon, walking slowly toward them. Wearing a Panama hat, ice ax, and big smile, the final member of the Manchester field team, Joe Macquaker, had arrived. He would continue the sedimentology analysis that Taylor and the LiDAR team had started. The LiDAR team finished packing up and left site for the last time, on their way to another canyon to scan in Utah. Macquaker corralled Chris Ott and headed toward an outcrop of Hell Creek. He was soon burrowing through the Badlands like a man possessed, with occasional squeals of pleasure as something interesting came to light. Macquaker is one of the most enthusiastic people in the world to talk about . . . mud! You can't help but agree with him, given his infectious laugh and passion for all things muddy. He and Ott wasted no time in trenching tens of meters of badlands in and around the site to investigate the past sedimentological history. The data collected from this work would help us to understand how and why we had mummified skin impressions around our hadrosaur.

Once the body block had been covered to its base, we were all very relieved, but we soon remembered that we still had to plaster the underside. Tyler and Tom Tucker had long been preparing for this particular part of the dig. They had devised a cunning way to feed long strips of plaster-soaked burlap through Norman's tunnels to wrap around the body block. A combination of purpose-made and modified tools allowed the burlap strips to be fed underneath the mummy, with a team on either side—again including "Hairy Dan" and me on one side with Tyler, Tom Tucker, and Steve Cohen on the other. Another MRF volunteer, Patty Kane-Vanni, got the fun job of chief plaster slop-and-goop maker, and she was soon covered head to foot.

We were hard-pressed to keep on schedule, because Tyler's brother would be arriving the following day to weld the steel frame beneath the body block. As the long strips of plaster-soaked burlap were skillfully pushed through Norman's tunnels, we gradually wrapped the hadrosaur mummy in its "security blanket." Lengths of 2x4 wood were cradled in plaster-soaked burlap and fed through the three tunnels which underlay the hadrosaur. Pulling each end of the burlap, we heaved the boards tight against the ceiling of the tunnels. These would serve as buffers between the steel frame and the body block. Several more lengths of burlap, gooped in plaster, were fed through the tunnels, ensuring the wooden supports were firmly held in place. By the end of the day we were all exhausted, but happy that the dig was on schedule. We had used nearly 400 pounds of plaster. Once the plastering was completed, we sat for ten minutes catching our breath in the dry heat of the Badlands. Each one of us had been drinking two gallons of water a day and we were still feeling thirsty.

One big question remained, though: Would the steel frame Tyler's brother had welded fit around and under our field jacket? Tyler's brother Derek and his colleague Johnny had constructed a steel box frame back in their workshop in Marmarth, and it had already been delivered to the site. Tyler suggested we put the steel frame in place to see if it fitted around the base of the body block. It took six of us to lift it into place, with an added complication. The steel frame had sat in the sun all day, and it was so hot that you could barely hold it without gloves. After a mutual chant of one, two, three, we lifted the frame over to the body block and gently slid the frame around the base. It did not fit, though. We had to make some slight adjustments to the base of the mummy block.

Fortunately, a plug of sediment on either side of Norman's tunnels lifted the hadrosaur above the floor of the excavated site. We had the roots of the body block to cut into and around to allow the frame to drop into place. After several tries, we finally got the frame to fit. Like an open square of steel it sat on the three sides of the base of the block. One of the long steel sections, which would complete the four sides to the frame, still had to be welded in place by Derek and Johnny. After that three solid steel rods would join the longest sides of the box-welded frame.

The welding trucks arrived the following day. It was amazing to see how quickly and neatly the welds were done. Each weld had to be perfect, as a weakness in any of the joints could prove catastrophic when lifting the ten-ton block. Derek and Johnny steadily welded up each joint, then ground the weld down with a rotating steel brush until each joint was smooth to touch. As they worked away on the frame, Ott and Macquaker were busy trenching nearby, collecting more sediment samples. Every now and then you would here a loud Macquaker laugh, followed by an equally loud pat on Ott's back—another piece of the sediment puzzle had no doubt just dropped into place.

After a day of welding, the steel support frame was in place around the base of the hadrosaur block. As Tyler and I regarded the enormous plastered body block and heavy steel frame, we were both worried that it was a tad on the heavy side, but in this case, heavy was good. The hadrosaur mummy was too precious to risk. In two days' time, the front-end loader and low-loader would make the long journey to the site. Only then would we absolutely be sure that the frame, plaster jacket, and weeks of work had been successful.

We drove back from the site, exhausted. To the northwest, we spotted a brush fire, but another one looked like it was a lot closer to Marmarth. A severe lightning storm the night before had set a few fires burning across the Badlands, but none near our dig site. The fire that we saw ahead of us, toward Marmarth, started both Tyler and me to worrying. How could we access the site if a wall of fire was in the way?

As we made our way toward the main road west of Marmarth, the stench of burning brush filled the air. Two members of the dig crew, Dan Pepe and Doug Hanks, were volunteer firemen in Marmarth. Once on the main road, the air became thick with smoke, Dan and Doug raced off toward Marmarth. We caught up with them a few miles short of the town. To the north of the highway flames licked their way through the tinder-dry sagebrush, and a team of firefighters followed in hot pursuit, beating out the flames. Dan and Doug were in the thick of the action—paleontologists and part-time firefighters.

We headed back to Marmarth and Pastimes, the finest bar and eatery in Marmarth and the only bar and eatery in Marmarth. We were pleased that the fire looked under control, but we still worried that it might cut off access to the site for the extraction of the mummy. We sat anxiously in Pastimes, waiting for further news. An hour later, Dan and Doug rolled in. The fire was out—a close call for all of us. The fire had been started by a farm vehicle, they reported. Its hot exhaust had been enough to set the sagebrush alight. Tyler and I both realized that the exhausts on our field vehicles could do exactly the same. We would have to make sure it was only dust we were kicking up and not smoke.

The day before the extraction of the tail and body blocks by the front-end loader, Tyler and the MRF team increased the number of supports between the body block and the steel frame beneath it. The final connections between the body block and earth were cut away from beneath the block and the holes plastered shut. The mummy was sealed in a protective cocoon of burlap bandages and plaster. It was now a dinosaur mummy in more ways than one.

I awoke at 5:30 a.m. on the day of the big lift. I had slept little the night before. Today we were meeting a front-end loader that would lift the two blocks from the site to a loading area about half a mile away and the low-loader truck that would carry the tail and body blocks back to Marmarth. We set off for the site at seven. Everyone was glad to see the vehicles were ready to go. Tyler was riding in his father's pickup, because the whole family had turned out to see the extraction, even his 85-year-old grandfather. Tyler looked nervous, mirroring how I felt. He admitted to me quietly that he had had a bad dream the night before in which a ten-ton block of dinosaur bounced out of a front-end loader's bucket. We would soon see how well his dream predicted the future.

The huge low-loader was halfway to the site already, as was the enormous John Deere front-end loader. Its 12-foot-wide bucket could lift 30 tons. The John Deere rumbled down to the site with considerable ease. Seeing how big it was, Tyler and I both decided that we had to get ourselves one of these monster diggers. It would make clearing dig sites and building access roads a breeze.

The first task for the John Deere was to remove the four-ton tail block from the site. It was already set on a steel sledge, welded two years earlier. The tail block was gently scooped up. Soon the huge front-loader bucket swallowed the tail block in a perfect fit. Too perfect, I thought. How would the much bigger body block fit? Tyler and I exchanged slightly worried glances. We helped the front-end loader crew chain the tail block in place. We did not want the tail to bounce out of the bucket. The enormous vehicle slowly backed up to turn around. It was like an elephant trying to turn around in a telephone booth, but the driver knew exactly what he was doing.

As the John Deere faced the right direction, Tyler and I walked in front of the huge vehicle, kicking or hauling out of the way any large boulders that might affect the stability of the wheels. The procession arrived at the low-loader and we watched as the bucket gently lowered the tail onto the back of the truck. It went off without a hitch, only a few splinters beneath fingernails, painful but easily forgotten. Grinning, the driver of the John Deere asked if Tyler, several of the MRF crew, and I wanted to ride back to the site in the bucket. He did not have to ask twice. As we gently bounced our way over the Badlands in the front-end loader's bucket, the hadrosaur body block loomed ever larger in the distance. It looked big, very big.

The John Deere smoothly lowered its bucket and we stepped back onto terra firma. In front of us rose the massive hadrosaur body block—I swear it had grown in the past hour. Everyone's eyes glanced at the bucket, then the enormous mountain of plaster and steel. We were all thinking the same thing. As the John Deere nudged toward the block, inch by inch, we began to breathe a sigh of relief. We had about three inches to play with on either side of the body block. It would fit inside the bucket, barely!

However, how do you shift a ten-ton block of steel, plaster, and rock inside a front-end loader's bucket? You can't just scoop it up . . . or can you? Rance Lyson, Tyler's father, stepped forward to help with the tricky maneuvering of the body block. Many years of working on the oil fields had provided him with practical skills matched by only a few. We were very glad to have him aboard.

The leading edge of the bucket was nudged beneath the long east edge of the steel frame. As the bucket was slowly and skillfully pushed deeper beneath the body block, the structure moved—for the first time in more than 65 million years. As the body block inched into the bucket, we saw a slight problem. The body block was so tall, it would collide with the top of the bucket before it was snugly inside. We would have to secure the huge block where it was, not fully inside the bucket, but balancing on the leading edge. All I could think about was Tyler's dream from the night before.

The lifting crew and Rance set about chaining the body block's steel frame to the bucket of the John Deere. We could not attach anything to the plaster, since it was there to support the fossil, not the weight of the frame and body block. The block was soon strapped and chained securely into the bucket . . . we hoped.

As the bucket slowly lifted the huge weight, the block gradually rose from the ground. A loud cheer broke out from the onlookers. Then the block wobbled on the front edge of the bucket. The hadrosaur was certainly trying our nerves. Additional chains were skillfully added by Rance and the crew to support the front and back of the block, stopping any rocking motion of the block inside the bucket. If it moved only a few inches farther back in the bucket, it would crush the plaster field jacket and fossil within.

Once the block had been resecured, and Tyler and I reassured, the front-end loader began to slowly move in reverse from the site. The John Deere deftly maneuvered up the steep slope of the butte that formed the north wall of the site. Once again, Tyler and I nervously walked backward in front of the John Deere, watching every shake or judder. Inch by inch we slowly led the vehicle with its precious load from the site. Unexpectedly, the right tire of the John Deere hit a large boulder, hidden from the view of the driver. The whole rig shuddered and the chains suddenly rang under the tension. I looked on helplessly as the whole body block shifted forward, but the chains held.

When the John Deere reached the relative flatness above the site, we all knew the worst part of the extraction was over. We kept ahead of the front-end loader, marching up to the low-loader. The last few hundred yards went smoothly, with both bucket and load behaving. However, I saw angry thunderclouds to the north threatening a new potential problem. If we had a downpour now, the first in months, it would mire every vehicle and person on-site.

As we arrived at the low-loader, the mood among the MRF and Manchester team was triumphant. With its powerful hydraulics the John Deere lowered its precious cargo onto the flatbed. The truck quaked as the full weight of the mummy was brought to bear on the suspension. The load was safely strapped and chained to the low-loader, the oddest cargo it had ever carried. Our strange convoy headed off toward civilization.

When the blocks had been safely delivered back to Marmarth, we headed off to Pastimes to celebrate. As we sat in the bar, the rain started hammering on the roof. We had just made it!

Macquaker, Ott, and I were leaving the next day. The operation had been completed without a hitch. Tyler's MRF team were an inspirational bunch to work with, as of course was Tyler. The excavation of any dinosaur is a dynamic process, as there is not a set procedure. Each one is different and poses its own problems to solve. Tyler is one of life's paleo-problem solvers and he has a great staff at the MRF to make things happen.

The next step was to study the fossil, sediment samples, LiDAR data, and other information gathered at the site. As we began to sift through the huge amounts of data and start to prepare the encasing layer of sediments, I was struck by a happy question: Who knows what amazing secrets were locked inside the encasing rock waiting to be discovered in the lab? The tantalizing skin impressions seen on so many parts of the fossil already made this find special. The bizarre soft-tissue structures in the cross-section of the tail now had to be properly analyzed and prepared before we could fully understand their significance. We still had a lot of work to do. While the excavation had ended, the story of our dinosaur mummy had only just begun.

When the giant field-jacket dinosaur was delivered to the lab, it was decided to flip the blocks in order to access the encasing matrix and prepare them from the underside first. The process of removing the encasing matrix requires skill and extreme caution. This is one job that cannot be hurried. Once the blocks had been carefully turned, using heavy lifting gear, the upper surface of the protective field jacket was removed. This was the first time we had seen the sandstones and mudstones on which Dakota had lain for the past 65 millions years. The paleo-lab was where the real detective work would begin, as every shred of information from the huge fossil would be recorded. We prepared to measure and collect samples as we removed the matrix, getting closer to the skin. The next stage of Dakota's long journey had begun.

CHAPTER EIGHT
THE CHEMISTRY OF DEATH

"I consider nature a vast chemical laboratory in which all kinds of composition and decompositions are formed."
—Antoine-Laurent Lavoisier

DAKOTA PRESENTED MANY PROBLEMS for the science team to solve, not least of which was how such remarkable preservation had occurred. To lock an animal in such a state, by-products of decay had to balance the preservation process. The balance commonly occurs when most soft tissues have been "chewed up" by microbial activity, freeing enough organic carbon, iron, and other decay products to preserve more resistant skeletal elements. Here, though, was a paradox: we had a skin envelope, so where did the materials to preserve our dinosaur come from? How do you keep a carcass from collapsing? To help explain this, let me use the example of a dead cow I once found out on the prairie.

The cow carcass looked in a sorry state; it had been dead a few weeks. The stomach contents had spilled around the carcass, yielding a halo. Coyotes had devoured the animal's rear end, consuming soft flesh from the inside out rather than chewing through tough hide. Loose skin draped over the uncollapsed skeleton like a poorly fitting shroud. The landscape and nearby scavengers were in the process of recycling the carcass. A month later, all that remained were bleached white bones scattered around a wide area and a lush island of vegetation where the carcass had once lain. This was one animal that was never going to make it into the fossil record.

The decomposing remains of plants and animals provide a high-quality source of essential molecules of life, including carbon and nutrients (e.g., nitrogen and phosphorus). Small animals are usually dispatched quickly with few remains left to mark the landscape, given most will have been processed in the digestive tracts of predators. Larger animals take more reprocessing and can make a significant impact on the microenvironment associated with a cadaver. The death of a large vertebrate in many environments can provide an oasis of energy in an otherwise low-energy system. Many species' life cycles are inextricably tied to death and reprocessing. Before we can understand the grave secrets of our mummy, we will have to take an unpleasant journey into the world of forensic taphonomy.

The decomposition of animals in terrestrial environments has not received that much attention, since most studies delve into the litter of plant life. However, animals, especially large ones, leave sizable bodies to decompose and be recycled in the environment in which they perish. Several tons of dinosaur would have been a significant feast for the organisms whose life cycles depended upon such mortality.

Ivan Efremov (1907-1972) can be considered the father of taphonomy, since his groundbreaking work in 1940 invented this specific approach to paleontology. Most studies since then have been undertaken by organizations like the FBI, who have a keen interest in the grave secrets of humans, especially those who end up in shallow graves as a result of foul play. Diagnosing the length of time and time of year that a body was buried can be crucial to solving a grisly crime. The temperature, moisture content, and insect activity are but a few of the variables that have to be considered when deciphering the decomposition history of a body.

The soil and microbial communities it contains are also critical to understand if we are to disentangle the taphonomic tale of a cadaver. Almost all terrestrial plants and animals end up in a patch of soil, whether that be in the parch-baked sands of a desert or the sodden channel sands of a river. We have to examine the above- and belowground ecology and microbiology to identify the key players in the recycling process. Many studies have shown that the melding of large bodies into soil is primarily regulated by the size of the body and the activity of scavengers and humble insects. Surprisingly, insects are key players in processing some of the largest animals that walk on the earth's surface today. Our taxonomic friend Linnaeus from the 18th century commented on this fact, saying, "Three flies could consume a horse cadaver as rapidly as a lion." More recent work has supported this view, indicating that insects can consume a body before a scavenger has fully utilized it. The complex intertwining roles of microbes, insects, and scavengers are also affected by season, for some species are more active at specific times of year. A road-kill deer can lie partially chewed all winter, with occasional scavengers taking a bite. However, the same deer at the height of summer would soon be consumed by the more active insect population. When one or more of the decomposition processes is inhibited, a cadaver can persist for much longer on a surface.

The study of decaying bodies has provided useful data on the various stages of decomposition that might also give clues to the time and type of burial environment. It is therefore worth spending a bit of time reviewing this ghoulish area. The best place to start is a fresh body of a vertebrate, say our hapless cow I met in Montana. Remarkably, as soon as its heart stopped beating, the oxygen that was so critical to life processes ceased to circulate and internal anoxic conditions rapidly developed. In the absence of oxygen, cells began their auto-destruct sequence, called autolysis. The cells self-destructed courtesy of enzymatic digestion. While the tissues of the cow were still relatively soft, prior to rigor mortis setting in, blow- and flesh flies all colonized the corpse with their eggs that soon spawned and became voracious larvae (maggots). All this occurs when the cow has been dead only ten minutes.

If no scavengers appear to pick at our dead cow, the fly maggots will soon compensate for this lack of attention. Within 24 hours the inside hide of the cow is filled with the writhing bodies of maggots, feasting on the rotting flesh. While the maggots are munching from within the carcass, soil microbes start the process of multiplying on the new source of food beginning to ooze through various orifices of the cow into the underlying soil. Microbial communities that were once symbiotic with the animal in life, living in the respiratory system, gut, and intestine of the cow, also begin to multiply and consume their host. The organic acids and gases (including methane and hydrogen sulfide) generated as by-products of the microbes' metabolism soon start to change the color and to bloat the carcass, accompanied by a distinctive rotting smell.

During the bloated stage, the structural integrity of the skin is usually maintained, unless the animal in question had wounds or punctures to its skin envelope from predators or scavengers. At this stage of decomposition the internal pressure generated by the gas inflating the body cavity forces decay fluids from the mouth, nose, and backside of the cow, a largesse welcomed by the soil microbes. However, eventually the skin of the cow ruptures at some weak point, often as a result of dining maggots. More flies can then get into the carcass to lay even more eggs, but oxygen-loving bacteria also join in the act of decomposition. We are now entering the most vigorous time in the recycling of the poor cow. The microbial metropolis is in full swing.

The combined efforts of insects, microbes, and by now scavengers attracted by the stench of death are making short work of the carcass. Most of the body fluids have now been released into the surrounding soils, causing an initial die-off of surrounding vegetation, due to nitrogen toxicity, leaving an almost scorched halo around the body. The maggots begin to pupate, having eaten their full. Depending on the soil porosity and permeability beneath the body, the sediments are directly affected by the decay juices of the body. As the soil microbes process the influx of nutrients from the carcass, their metabolic by-products also begin to alter pore-water chemistry (the water filling the spaces between grains of sediment), giving rise to the precipitation of early mineral cements to bind the soil particles. The body fluids of carcasses have been shown to increase the pH of soils, and with large quantities of organic carbon present, carbonate cements might rapidly form. However, for this to directly affect the preservation of the animal, it would have to be at least partially buried in the soil. As the ground immediately around our cow receives a pulse of organic carbon and nitrogen from the carcass, again the soil chemistry and microbial community have to manage the influx. If the soil is relatively anoxic, the infusion of carbon can result in rapid mineralization. This is good news if the carcass is buried and we want it to become a fossil. However, not only carbon is important at this stage. The bodies of vertebrates are a wonderful store of elements that, once released from their biological bonds, can form different species of mineral, but again here is the paradox. Release too many of the organic building blocks to the inorganic processes and you reduce the amount of organic remains to be fossilized.

At this stage of decay, our cow is a now disjointed patch of skin and collapsed skeleton in a pupae-rich soil. Most of the immediately available nutrients and moisture have been sapped from the ill-fated cow. It is time for the microenvironment that has been deluged by the carcass to regain control and impose wider conditions and controls on the patch of earth on which the remains lie. This process can take years. Studies on the decay of large vertebrates, such as muskox (*Ovibos moschats*) and bison (*Bos bison*), have shown effects on local plant communities that last five to ten years. This suggests the soil microbial community might also be affected in the long term by the enhanced nutrient levels relative to surrounding ground. However, any gardener will tell you to "plant" your deceased pets beneath your favorite rose bush!

So far we have only focused on surface processes. What happens when you bury a body? All of the remains that we find locked in the fossil record have been subject to burial at some point in their history. The fossils of Ludlow and the Burgess Shale have shown that rapid burial can enhance an organism's preservation potential. However, the fossils of Solenhofen and Santana indicate that the depositional environment can also inhibit microbial decay or scavenging, also enhancing your preservation potential. How important is rapid burial and the chemistry of the grave soil?

Rapid burial restricts access to most insects and scavengers, resulting in much slower decomposition. If burial occurs in coarse soils prone to losing moisture rapidly (e.g., desert sands), the hydrolytic enzymes associated with the decomposition processes can be retarded. This is what helped preserve many of the predynastic Egyptian mummies. However, if moisture content is elevated in such coarse-grained soils, a moldlike image of soft tissues is often persevered in the soil. Such preservation was observed in the Anglo-Saxon (sixth and early seventh century) Sutton Hoo burial site in Britain. Burial in fine-grained (clay) soils, especially when wet, can also result in slower decomposition of cadavers, possibly due to metabolic constraints placed on microbes in such environments. As a result, the decomposition is left to less efficient anaerobic microbes.

If a body is buried in an environment where reducing conditions persist, adipocere may form, significantly slowing the decomposition processes further. Adipocere, or "grave-wax," is a grayish fatty substance formed in grave soils that are often saturated with alkaline pore waters. The resultant body can resist further decay for hundreds of years, often preserving facial detail and even skin wounds. This poses a major problem for many graveyards that wish to reuse grave soils for burial, since they are prevented from doing so by the preserved bodies. In some places in Europe, such as southern Germany, 30-40 percent of graveyards are affected by this problem. For a paleontologist such a phenomenon might aid the preservation of trace fossils that might survive long enough to leave their mark in the fossil record. However, once exposed to surface processes, adipocere will continue to decompose.

As the science of forensic taphonomy shows, many things can be learned from grave soils, but what of the many sediment samples we collected during the excavation of Dakota? Might they hold some of the elusive grave secrets of our dinosaur?

PARADOXICAL DINOSAUR

The fossil's soft-tissue structures preserved in high relief in what appeared to be the mineral siderite (iron carbonate, $FeCO_3$) presented a great scientific opportunity to investigate both the paleo-environmental conditions under which it was preserved and specifically the dinosaur mummification process.

The fossil raised a number of very intriguing scientific questions. First, how could such a large organism actually be mineralized with siderite? More particularly, how could sufficient iron be transported rapidly enough to the decaying organism to preserve its soft tissues? If the process was too slow, the skin, muscle tissue, etc. would not be preserved in 3-D. This fact alone implies that large volumes of either available mineralized iron were on hand in the sediment, or significant concentrations of reduced iron existed in the pore waters at the time the decaying dinosaur was being preserved. Second, why did an iron-reduction (presumably fostered by bacteria) degradation process occur at this site rather than the more common oxic degradation processes—in other words, what was the paleoenvironment just after the dinosaur died?

The early organic decay processes in sediments are usually driven by the availability of oxygen (the oxidizing agent). Oxygen typically is diffused in the pore waters from the overlying atmosphere. Here, however, oxic-degradation processes preserving the dinosaur are not an option because during oxic degradation all the soft tissue would have been eaten. Plus, reduced iron would not have been available (this is significant because the dinosaur was preserved as siderite, which contains reduced iron). In terrestrial environments, in addition to oxygen, oxidized iron (an alternative oxidizing agent) can also drive the organic matter degradation processes, particularly when the sediment is waterlogged. Iron, however, does not diffuse as a gas into the sediment in the same manner as oxygen, but rather is buried with the sediment as a solid, such as an iron-oxide coating on sand grains. Given that the dinosaur was preserved in a freshwater setting, methanogenesis seems the most likely decay option. Methanogenesis is a process driven by microbes known as methanogens. Such microbial organisms are capable of producing methane as a by-product of their metabolism, in the same way we produce carbon dioxide as we exhale. This process usually marks the final decay associated with organic remains. The paradox in this case is the need to reduce the iron in the first place, then transport sufficient iron to the mummification site to rapidly preserve a very large volume of soft tissue, and finally to ensure that bicarbonate from the decaying animal was mineralized rapidly (i.e., before the carcass collapsed). When close to equal amounts of carbonate ions and carbonic acid (from organic material in groundwaters) are present, it can form a buffer system, helping to maintain a constant pH that is ideal for the precipitation of specific minerals.

Given all the special conditions required for dinosaur preservation, it is not surprising that such dinosaur fossils are rare. With this animal, however, we had a fabulous opportunity to investigate all these processes. So our scientific program sought to address how dinosaur preservation might have been optimized to form this extraordinary mummy. We set out to investigate the fundamental ancient geochemical processes that caused the soft parts of this dinosaur to be mineralized.

The samples of rock that we collected in and around the mummy site needed to be analyzed. Each grain had the potential to reveal information about the environment that existed before, during, and after Dakota was alive. As sands and muds accumulated along the banks and beds of rivers, they often would have become bound by specific cements that related to the chemistry of the waters flowing above, below, in, and around Dakota's resting place. Each sample would have to be carefully prepared and analyzed in an apparatus that would identify specific chemical signatures that would help reconstruct a picture of the prehistoric landscape. During the dig Macquaker and I had collected hundreds of samples; some were taken close to the skin, others from what looked like a fossil riverbed in which Dakota's body came to rest. All were bagged, tagged, mapped, and recorded.

The type or types of cement binding the rock and the different types of grains that the rock was composed of had a major effect on the preservation of Dakota. To understand this relationship, we had to bring the samples back to the lab so that we could start preparing them for the many analytical techniques we were going to employ at the University of Manchester.

HEAVY ROCKS, ELECTRONS, AND X-RAYS

The two sports bags looked full. The airline check-in clerk leaned forward to lift the bag off the scale, not having paid attention to its weight. He did a double-take, checking first that he had not snagged the bag on the mechanism. Looking up at the digital readout for the bags' weight, he shook his head. "You're over your luggage weight limit, sir." Shipping the sediment samples back to Manchester would have lost us valuable time, so I bit my lip and paid the excess baggage. The sediment samples that had been collected over the last month's field work disappeared to be security checked. I could only wonder what they would make of a bag of rocks!

As soon as we returned to the University of Manchester, the bag of sediment samples had to be logged by Macquaker and sent down to the thin-section prep lab. There each sample would be sliced and ground into a suitable format for the particular apparatus in which it would be analyzed. This was a huge batch to process, and I left the fun task of explaining what was wanted to Macquaker.

Harry Williams and Stephen Stockley work in the thin-section prep lab of the School of Earth, Atmospheric, and Environmental Sciences. They had to slice samples of dinosaur sediment, skin, tendon, and bone for the inorganic study components of the project and were assigned to make hundreds of polished thin sections of the sediment samples. A thin section consists of a thin sliver cut from each sample with a diamond saw, mounted on a glass microscope slide with a strong resin, and then ground smooth using progressively finer abrasive grit until the sample is only 0.03 mm (30 micrometers) thick. The skill needed to make any thin section is difficult to explain, for the technician almost needs a feel for the material being sectioned. It is easy to abrade right through a sample. The process is both fiddly and time-consuming, taking many hours of concentration for each sample. The final product is a wafer-thin slice of rock, so thin that light can be transmitted through it.

The thin sections can then be viewed on both optical and scanning electron microscopes, not to mention a number of x-ray methods. When placed on an optical microscope between two polarizing filters set at right angles to each other, the optical properties of the minerals in the thin section alter the color and intensity of the light transmitted through the section. As different minerals have distinct optical properties, the type can be easily identified. We hoped this work would reveal the origin and evolution of the parent rock that was above, within, and below the mummy site.

In addition, Tyler had given us a small sample of skin from the base of the tail block. We had to really think about how this precious sample would be used. It had not been touched by any consolidant (glues and resins used to bind the fossil) and the skin was still sandwiched in a layer of matrix. This was perfect, as we would be using samples from this same skin block in the hunt for biomolecules in the fossil skin. After Williams had made the first thin section of the edge of this sample, we realized how deep the "skin" layer of the fossil was. It raised our hopes that organic matter might be locked inside the minerals that had, with a bit of luck, not completely replaced the skin.

The School of Earth, Atmospheric, and Environmental Sciences has access to some of the best analytical and imaging facilities in the United Kingdom. These were to be liberally applied in many facets of the mummy project. The first port of call for the thin section of skin was the department's Environmental Scanning Electron Microscope (ESEM). Most people are familiar with high-resolution SEM images, which have an almost 3-D quality. Typical examples show an insect's compound eyes, the hairs on the legs of ants, etc. Conventional SEM requires samples to be imaged in a high vacuum, and the sample coated in either carbon or gold. With ESEM the sample can be imaged "wet," requiring no coating, and run at a lower vacuum. Scanning electron microscopes are capable of producing incredibly high-resolution, almost 3-D quality surface images of samples. The image is produced as a function of the energy exchange between the electron beam and the sample, resulting in the emission of electrons and electromagnetic radiation.

I had first cut my SEM teeth at Manchester 17 years earlier while working on fossil eurypterid respiratory structures. However, the department did not own their shiny new ESEM at that point, and I now had to learn how to drive the new machine. Fortunately, the department has Steve Caldwell to hold our hands. We both looked at the polished thin section of the skin and carefully marked it with silver solution to make it easier to navigate around when under the ESEM. Caldwell opened the main sample chamber of the microscope and carefully mounted the slide onto the imaging stage. After making sure the slide was centered, he closed up the chamber and switched on the pump to evacuate the chamber to an operational vacuum. Caldwell turned on the power to the filament, the source of electrons that would be focused into a tight beam onto the sample. Before long, we were examining the surface of the thin section.

After several hours of using the ESEM monitors, I was becoming proficient at "flying" over the surface of the slide, mapping and imaging structures on its surface. The skin layer was distinct from the surrounding matrix, much finer grained. The ESEM also possessed a clever tool in addition to its imaging capabilities, Energy Dispersive X-ray Analysis (EDX). EDX works on the principle that each element of the periodic table has a unique electronic structure. When an electron beam is aimed at a sample, each element provides a unique response to the electromagnetic waves, releasing information from each atom in the form of an x-ray. The analysis of the x-ray emissions helps characterize the specific elements present at the spot being targeted.

The ESEM images were very surprising. On the monitor I could clearly see the cross-section of the dinosaur skin. This I expected, but the minerals that made up the skin were very distinct from those of the surrounding rock matrix. For the first time I knew we had uncovered not merely a skin impression, but a mineral replacement of the skin. This is a big difference, because we were no longer dealing with a fossil impression; this was fossilized skin!

As I skimmed over the surface of the slide with the ESEM, the details became clearer. The skin apparently had been replaced with a combination of fine clay minerals and the ubiquitous siderite. I was proceeding cautiously, though. As is the case with many fossils, you sometimes see what you want to see. Could the strange globular structures be collapsed cells? Could the distinct layers I saw in some parts of the skin envelope represent original layers within the dinosaur's skin? One question truly cried out to be answered: Could any biomolecules have survived inside the mineral matrix that had replaced the skin? Mary Schweitzer's work, as we will see later, has caused many paleontologists to revise their opinions about which molecules might possibly survive in the fossil record.

I needed to speak to three people in the department to confirm or disprove my possibly wishful thinking—Joe Macquaker,

Roy Wogelius, and Andy Gize. Macquaker was already plowing through the many sediment samples on another SEM unit in the department. His task was tracking down the geochemical signatures of the Hell Creek Formation environment prevailing before, during, and after the life of our dinosaur. Wogelius and Gize were not original members of the mummy team, but they possessed the crucial skills of inorganic and organic geochemistry, respectively. I contacted Wogelius from the Williamson Research Centre, within the department, and arranged to meet.

Over lunch, Wogelius and I talked through what I had viewed on the ESEM. He was excited by my preliminary findings, especially since dinosaur skin was not his usual line of research. He shares a joint position between the department in which I worked and chemistry, giving him access to a vast array of analytical techniques practiced in each department. He has spent many years developing one technique in particular, which involves technology with its roots in the U.S. military. This is fitting, since Wogelius originally hails from the United States. With his thoughtful gaze and gentle smile, he talked me though which organic and inorganic analysis routes we might take with the mummy samples. Although we might be able to recover an elusive biomolecule, Wogelius said that any such claim would have to be validated by several techniques and at least two different labs. We parted company, agreeing to exchange samples and ideas later in the week. In the meantime, Wogelius was going to talk to his colleague Andy Gize, who would undertake the initial hunt for organic molecules.

I checked in to see if Joe Macquaker could help confirm my ESEM findings. I found him sitting at the helm of the department's Jeol SEM. The unit was now nearly 20 years old, but Macquaker preferred it to the newer ESEM, claiming with a matter-of-fact grin, "It can do a lot more than the new unit." He swiveled his chair back to the monitor, where he scrutinized another slice of mud rock. "Where's your sample, Phil?"

Before long we were again flying over the monochrome microscopic surface canyons of the thin section. Macquaker expertly spun dials, pushed buttons, and guided the sample beneath the electron beam. The images that he was plucking from the slide's surface were quite remarkable. Soon we were firmly located within the skin envelope's cross-section, mapping the surface chemistry and structures. Small dark pockets of material, only a few microns across, pocked the surface of the skin. Macquaker focused on these pockets and hit them with the EDX analyzer. "There's your organics, Phillip," he said matter-of-factly. When I did a double take, he added, "Don't look so surprised. Any geologist will tell you that organics are far more common than most people think. How do you think your car got to work today?" Thinking about this for a second, I could not help but smile. He had touched upon the underappreciated fact that the entire world's economy hangs upon oil. These organic fossil remains had locked past energy from the sun to be released millions of years later by the internal combustion engine.

The organic material in the skin layer needed to be analyzed further by Wogelius and Gize, but at least we now knew the discovery had potential. But was it a contaminant? Only time would tell. In the meantime Macquaker carefully conducted sample-removal protocols for the thin section to reload a sample from the sediment that had once cloaked our dinosaur. After his hundreds of hours on optical and SEM work, Macquaker was ready to tease out the past environments locked within the samples we had taken the previous summer.

It is important to remember that floodplains are complicated environments composed of a mosaic of microenvironments, including soils, river channels, riverbanks, lakes, and swamps. Looking intently at another multifaceted slide of minerals, Macquaker explained that these environments are complex kitchens where many chemical reactions occur because the building blocks of life—namely food, nutrients, and water—are present and locally abundant. For instance, food to support life is present because plants commonly grow on the floodplains. Oxygen, the molecule that most herbivores use to consume their food, is also available from the atmosphere. Water, which is used to transport food and waste products within organisms, is also available in the rivers, ponds, and lakes.

He went on to explain that the cycle of life that occurs on floodplains, however, does not just involve organisms, water, and oxygen. Also to be considered are the sediments in the soils—varied mixtures of weathered underlying rock, organic matter, and newly formed minerals. Plus, water would have been present between the grains in the sediment, or pore spaces. This complexity arises because when organisms die they are commonly buried by natural processes, including floods and the bodies of other organisms, such as plants. The presence of organic matter in the sediment can have a profound effect on the reactions that occur during burial. If the soil is waterlogged, the rate of organic matter degradation commonly is slowed. The resulting buildup of these decay products, particularly carbon dioxide and humic acids in the pore water, is much less acidic in streams, ponds, and lakes.

The existence of acidic groundwaters can have a profound effect on some of the minerals present in river sediments, particularly where they have been derived from rivers that drain volcanic highlands. These sediments commonly contain a large proportion of minerals (for example, feldspars and micas). Rock fragments that react with acidic groundwaters, and minerals that precipitate in direct contact with an oxic atmosphere (for instance, iron oxides), commonly cause the sediment in these settings to have a rusty hue. The reaction between these unstable minerals and the acidic groundwaters causes the feldspars and micas to dissolve, liberating elements such as magnesium and calcium into the pore waters. It also results in the precipitation of new clay minerals, such as kaolin. The dissolution of oxides and aluminum causes the pore waters to become enriched with iron, and the oxide coatings around the sediment are leached. The samples that

Macquaker was studying from in and around the mummy site were prime examples of this process.

Complicating the matter further is the fact that groundwater in the subsurface is commonly not static, but instead moves from areas of high runoff, such as rainfall occurring in high ground, to regions where the runoff is lower, particularly in lowland regions where the water table is lower. Macquaker was keen to explain that the movement of the water causes materials that have been dissolved in one area to be transported to other regions where the chemical conditions may be more favorable. Subsurface fluids migrate through porous sediment. In a floodplain the most porous sections are typically the sands associated with river channels. These channels thus become sites where minerals tend to dissolve and precipitate again. That's because, on the one hand, the fluids in these settings are continuously being replenished by acids from the decay of organic matter and humic materials, and on the other, the resulting dissolved products are constantly being removed. Our mummy was set in such a sand environment. The pieces of the sedimentary jigsaw puzzle were slowing falling into place, thanks to the work of Macquaker.

What did this all mean for the mummy? The decaying dinosaur carcass, which represented a large volume of reactive organic matter, may have profoundly affected local groundwater composition and the reactions that followed as it created a so-called cadaver decay island. The decaying dinosaur, as in the case of our cow on the plains, would have introduced a great deal of bio-available carbon and nutrients (particularly nitrogen) to the local environment. Decomposition of soft tissue from a dinosaur, especially in anoxic environments, might have caused the pH to rise, due to the work of ammonia-producing bacteria. If adipocere, or grave wax, formed, it would have mixed with alkaline pore waters and iron and calcium in solution, due to weathering. That might have led to partial mineralization of the animal's skin and fossilization of soft tissue, causing "mummification" and enhanced organic preservation. This was Macquaker's working hypothesis, one that was supported by the evidence that he painstakingly teased from the samples we had collected during the excavation.

Thankfully, the LiDAR survey data that Dave Hodgetts had just finished processing provided powerful supporting evidence of the waterlogged soils that Macquaker's hypothesis predicted from the mineralogy. A stone's throw away from the mummy site was a "fossil" channel of a huge river. The 3-D digital outcrop map was able to clearly mark the meandering channel, which had come perilously close to the burial site. That channel would have existed very soon after our animal was interred. The presence of this large watercourse would have ensured that the porous sandy sediments encasing the mummy were constantly bathed in the organic-rich waters filtering through the floodplain sediments. However, if the meander of the channel had cut a few feet to the east of its course, the mummy would have been swept into the channel and its remains pulled apart and redeposited as isolated bones downstream.

STABLE OR UNSTABLE?

To understand a little more about the floodplain environment required the special skills of Jim Marshall at the University of Liverpool. He was working on samples of the carbonate concretion that Macquaker and I had collected from around Dakota. Marshall was subjecting the samples to both stable isotope analysis and a special kind of microscopy. This would allow the team to look deep inside the environment that prevailed when Dakota was buried.

For the stable-isotope analysis, conducted inside a vacuum chamber, he dissolved in acid the carbonate minerals that had locally cemented the sandstones around the corpse. He then analyzed the carbon dioxide freed by the acid attack. Marshall had a very good reason for subjecting the samples to this analysis, as he wanted to understand which carbon isotopes were present.

Isotopes of an element are atoms with different masses—with different numbers of neutrons in their nucleus. For example, carbon has three common isotopes. Carbon with mass 12, and 12 corresponding neutrons, is the most common isotope and accounts for about 99 percent of all naturally occurring carbon—in you, me, and in rocks. Carbon with mass 14 is well known because it is used for dating, but it was of no relevance to the study of Dakota. Carbon-14 is formed in the upper atmosphere where the sun's radiation knocks bits out of nitrogen and converts it into carbon, but radioactive carbon-14 is converted into boring carbon-12 quite quickly. Carbon-14 is therefore useful for dating archaeological samples, which are a few thousand years old, but any carbon-14 in Dakota's skin would have decayed to below detectable levels shortly after the dinosaur died, in a fraction of the first million years.

This leaves us with the other carbon isotope, a carbon atom with a mass of 13. Carbon-13 is present in about one percent of all natural carbon, but it has the great advantage of being stable. Marshall knew this isotope could be used as a "fingerprint," even in sediments as old as those that surround Dakota. The proportion of carbon-13 in comparison to carbon-12 in minerals could be measured with some accuracy. This was important because the two types of carbon acted as a clue that could tell us where the carbon came from and thus how Dakota was preserved.

Marshall analyzed the carbon isotopes in the carbonate minerals—both in the calcite and the siderite—around Dakota's skeleton to trace its source. The microbes living in and around the rotting corpse would have ingested organic matter and soil-derived carbon. These microbes break down the carbon sources and generate bicarbonate, which we also find in the minerals. Marshall's preliminary results demonstrated that more than one source of carbon was available to the growing minerals, but that methane-generating forms were definitely involved. So once again the methanogen microbes reared their "heads" in the mummification process.

At the same time Marshall was analyzing the carbon isotopes, he also examined the proportions of different oxygen isotopes in the minerals. This study gave him a different sort of fingerprint, but one that is nonetheless useful in finding out how and where Dakota came to be preserved. Oxygen isotopes tell us about the source of water—rainwater, for example, has a very different isotopic signal from seawater—and this difference is influenced by where you are in the world (latitude and altitude). The oxygen isotope from the cements surrounding Dakota therefore might tell us about what sort of groundwater was involved and whether that water changed as the minerals grew. One possibility is that the preservation of the carcass would have been enhanced in groundwater that had become saltier, perhaps through evaporation in the soil. In the minerals analyzed by Marshall so far, however, the relatively light oxygen isotopes demonstrate that the minerals grew in normal rainwater, with little or no evaporation and no influence of the sea. Other processes must have caused the minerals to grow soon after the animal died, and preserved the shape of the skin envelope.

SOMETHING GLOWING IN THE DARK!

Following the electron microscope studies of the rock samples at Manchester, Marshall and Macquaker decided to take a closer look at the texture of the mineral cements that bound each grain of the rock together, again using apparatus at the University of Liverpool. Marshall suggested a technique that could literally shed light on how or why the minerals had formed.

The technique he proposed using was cathodoluminescence. This involves hitting the surface of a rock sample with a low-energy, diffuse electron beam. This treatment is very different from the highly focused and controlled beam used in the SEM. The energy from the electron beam is absorbed by the minerals, causing them to glow, though not at all brightly. Different minerals with different concentrations of trace elements and various ranges of defects in their crystal structure glow different colors.

For cathodoluminescence you need to sit patiently with a microscope—in the dark!

The cathodoluminescence work undertaken by Marshall looks like it will prove to be very productive. In one sample that may include part of the mummy skin envelope, the mineral cements show a tantalizing pattern. When Marshall studied the pattern, he realized that what he was examining might reflect the finer details of skin texture. This was exciting news for Wogelius back in Manchester, since he and I had already suggested the same thing from the ESEM work. It's a great feeling when independent work and methods draw similar conclusions. However, as in most scientific breakthroughs, we need clarification of such a wild possibility. The exciting isotope and illuminating cathodoluminescence work continues.

The study and interpretation of such samples can take months or even years, but the work we have already completed tells us many things, especially when combined with clues from other fossils. The fossil remains of the plant material that Tyler found during the initial excavation tells us much about the environment, since specific plants make excellent climate indicators. The presence of plant material in many of the rock samples collected might also give clues to why siderite and calcium carbonate cements were present. When plant material is abundant in waterlogged sand, mud, and other sediments, its presence can increase the acidity of the groundwaters. This helped many chemical processes to start very early after Dakota's burial. The early formation of specific cements, such as siderite, might have kick-started the journey for Dakota to become a mummy.

Other clues need to be studied as well. Our hair is different from our skin, our skin from bone, and bone from muscle, and the same can be said for any animal. Each body part has its own specific structure and chemistry. When buried, a body will react in different ways to the environment in which it is situated due to this variation in body chemistry and also to any variations that exist within the sands or mud in which it is encased. The absence or presence of groundwater, changing due to seasonal fluctuations in climate, would also have affected the mummification process. The season might also have influenced the level of insect larvae infestation, given the variation in life cycles at different times of year for different species.

The siderite cement that surrounds much of Dakota was also concentrated in the sandstones and mudstones that immediately surrounded the fossil, especially below. When we lifted the body block from the excavation site, the dig team all commented on the strange iron-rich minerals staining the ground beneath the body. This was not fossil blood, but related to the slow seepage of Dakota's body fluids that eventually sank into the sands and mud beneath his body, in the same way that water sinks through sand on a beach. As the body fluids seeped into the sediments, the microbial soil community would have taken advantage of the sudden pulse of nutrients from the mummy. Resulting byproducts of the microbe metabolisms would have affected pore-water chemistry and given rise to the complex cement colors we observed beneath the mummy's body.

Earlier we mentioned methanogenesis and how it might have given rise to the critical pore-water conditions suitable for the precipitation of preserving minerals. The gut of Dakota, as with modern animals, possibly contained anaerobic microbes, including methanogens. These microbes help to break down cellulose in foods into digestible forms. Without such microbes in their stomachs, cows would not be able to process grass. The methane by-product is released by belching or flatulence.

The next vital step was examining the clues from Dakota's skin itself. This layer was very, very delicate. Although it had mostly been replaced with siderite, the skin was still much softer than most of the surrounding sediments. Separating the rest of the matrix from the skin was like trying to remove a suit of armor with a jackhammer while trying not to damage the occupant! The tools we use to remove the excess particles have been often complicated by the tough siderite cement (Dakota's suit of armor) that Macquaker, Gawthorpe, and Taylor had spotted in the field. The preparators remove the cement-bound rock, grain-by-grain, using pneumatic micro-air pens and even dental tools to pick away the debris from the skin. The preparation of Dakota was, and continues to be, a long story of dust, grime, frustration, and unexpected surprises. The preparation process takes thousands of hours and immense amounts of patience. This is one job that cannot be rushed.

UNWRAPPING DAKOTA

While many were busy in Manchester, the folks back at the Black Hills Institute (BHI) were preparing the tail and body block. First they had flipped the block over, no mean feat. Then they had to cut away the steel support frame that Tyler's brother had constructed. Next, they built two giant "skateboard" supports that would allow the blocks to be more easily maneuvered around the prep lab. When this was ready, the two huge blocks were wheeled into the lab.

Tyler and I had thought long and hard about where to start with the mummy. We both decided that the tail block was best, because the preparators would be able to see the skin layer clearly and develop a good feel for the matrix, bone, skin relationships, and properties of various materials.

While the BHI were busy with the tail and body blocks, the arm that had been removed from the mummy the prior field season was already well on the way down the prep road with Stephen Begin, who was based in Michigan. Begin is, if he will forgive me for saying so, one of those distinct people in life whom you will never forget. He is usually clad in inappropriately tight shorts and a scruffy T-shirt, and his keen eyes analyze you through a pair of round-lens spectacles beneath close-cropped hair. His tanned hide betrays the Badlands excavation passion that he pursues each summer, helping Tyler run digs for the MRF.

He spent the summer of the mummy excavation a few ridges away extracting a *Triceratops* from an awkward precipice. While his skills were missed at the mummy site, the MRF volunteers at Begin's site benefited from his years of experience in the field. He is one of the most diligent, brilliant fossil preparators that I have ever met. Both Tyler and I on several occasions have wished that we could clone him.

Over the long winter months Begin, Tyler, and I remained in touch, exchanging e-mails and ideas. Both Tyler and I looked forward to the communications from Begin, accompanied by exquisite images of his recent work. His communications were usually entitled, "only a few new pix," and enumerated suspicious blobs and embedded concretions. A typical e-mail from Begin would also recount how hard the mummy material was to prepare. "These are very slow to prep because the farther up the valley I prep, the deeper they are embedded in the concretion and the harder it is to tell the difference between the two. How slow? I put in 14 hours last week, and most of it was spent on the scales on the distal side of the scale bar. That's maybe 100 square centimeters and 1 to 5-plus millimeters thick in about 10 hours!"

Begin had already made one trip to BHI to talk through his long experience of prepping the mummy. When asked if he could make another trip, his reply was typically brilliant: "Timing is the important part. I have a timber sale being cut for a client, so I cannot leave until they are done, at least another three weeks. I am also in the midst of beer brewing, batch two (lager) and three (ale), you know that real stuff, not kaylarada urine, so I need to factor in a switchover from the primary to the secondary fermentor and getting to the bottling point with the other." Here was a man who had his priorities right!

Through the cold winter months Begin pushed on with the prep. Soon the arm of the mummy appeared in full 3-D glory from his intense efforts. However, the prep was still tough going, "I have found several places where the skin layer goes into or under the white sandstone and both are relatively insane to try and prep for scale definition. The scales in the dark brown sandstone are also proving to be a royal pain in the arse. They do not prep nicely. Sometimes I am not sure if it is or is not skin because there are tiny bits of floating ironstone pebbles that are shiny just like scales. I have prepped a lot of them and you cannot tell what they are without being real careful." Tyler and I were both keeping our fingers crossed that we had not asked too much of Begin.

While prepping the mid-section of the arm, Begin was able to differentiate between a number of scale types along the length of the limb. The scales were beautifully preserved, or to quote Begin, " these babies are amazing. " His eloquent e-mails would talk of the "gently rolling hills and valleys" of the limb's surface as he separated the 65 million years of matrix. He continued prepping at least four hours every day, meaning the progress was visible at the end of each week.

Tom Tucker, who had been so critical to the excavation that summer, had taken the loose foot block of the mummy for preparation. He too was gently prising off, grain by grain, the tough matrix encasing the foot. Tyler had the tip of the mummy's tail in his lab at Yale. He had already removed much of the encasing matrix to reveal a stunning 3-D tip of the tail, complete with modified ridge scales running along the top, as seen on the back of the Sternberg and Senckenberg mummies. The BHI was also making headway with the huge body and tail blocks, and at a certain point Stephen Begin made the trip from Michigan to South Dakota to ascertain their progress.

When Begin arrived at the BHI, he did not set in to remove matrix too. Instead he reviewed prep techniques and new ideas with the BHI preparators John Carter and Debra Christie. They dedicated several days to examining a new type of skin matrix that had been exposed along the base of the tail on the main block. Unlike everywhere else, this skin appeared to have a moderately hard white sandstone rather than siderite base. A pinkish-orange bleed zone appeared between the white sandstone and the chocolate brown skin band, but it was very hard to tell if they had reached the skin until "Oops, bloody hell, you've gone through it." The skin band, now referred to by Begin as "the integument band," was less than 0.5 millimeter in places, so using the standard air scribes had to cease. The preparators were all reduced to using ultra-sharp steel and tungsten carbide points to pick individual grains of matrix from the blocks.

Another problem with the body block concerned what the prep folks at the BHI described as a section of folded, flappy skin. Bits and pieces really, but they were probably all connected. The problems they faced here were multifold. The section included a bit of pebble skin in which the pebbles were one- to two-millimeter ironstone dots and popped off, leaving craterlike ironstone impressions. The bands here were also thinner than usual, and the skin was very hard to see. There wasn't a bleed zone for warning when the skin would end or begin, just orangey-yellowy-brown sandstone and then, unexpectedly, more skin. Adding to the difficulty was the fact that the sandstone matrix was somewhat hard. Finally, the relative position of skin pieces to each other was so confusing that Begin was lost for words to explain. This complication concerned me; the mummy was not giving up its secrets willingly. It was beginning to look like the skin envelope did not continue onto the chest of the animal.

"You guys ought to be happy I don't have a family or any family members closer than 1,000 miles to visit and while away the holiday hours, so I spent a good hunk of time in the basement with my ol' buddy," said Begin, once back in Michigan. Yet he was making spectacular progress with his mission. At a glance, an uneducated eye might say it looked like a dinosaur's arm. To an educated eye, it looked simply incredible. He had unearthed a complete, unbroken layer of uninterrupted skin from the elbow to the fingertips.

Skin completely surrounds the bones of the arm and hand, stretching like a scaly envelope. The size and shape of scales vary across the arm, with possible bands of scales alternating in size near joints. This might have helped the tough hide of Dakota flex when he was alive. The bands of scales might also reflect color changes in Dakota's hide, but this is impossible to know. The hand is covered in skin, with delicate webbing joining the middle three fingers. These three fingers were surrounded in a mitten-like sack of skin with a beautifully preserved fleshy pad to walk on. However, the pad was too small for Dakota to have walked upon all day and was probably used only when browsing for low food, to steady his vast bulk, or when crouching to drink. The bones in the arms of *Edmontosaurus* have always looked too insubstantial to allow continuous use for walking or running, and Dakota's full arm seems to support this theory.

The broad webbed hands and large feet possibly helped him survive in more swampy environments or helped when crossing the large rivers that crisscrossed the giant Hell Creek floodplain. The first and fifth finger might have helped Dakota manipulate his food, grasping branches and tugging them toward his mouth, so his tough keratinous beak could cut the vegetation. Then it would pass back over his hundreds of teeth for grinding up the plants, to help digestion. We too chew our food, because this is a great way to increase the surface area of the food to the digestive juices (enzymes) that help break down and extract nutrients.

As more skin was exposed, Begin could recognize specific scale shapes and patterns, previously found only as isolated skin patches in other fossils. Like a jigsaw, Dakota was helping us position specific skin types on specific parts of the body. He did have that hole in his skin envelope, but even with the skin pattern from the chest missing, it might still be possible to fill the hole and make Dakota's skin envelope complete. What's more, the Sternberg mummy at the American Museum of Natural History has wonderful skin impressions on its chest, so it is the missing piece to our puzzle. For the rest of the body, Dakota has the potential to become a key for many hadrosaur skin types.

We had a hypothesis that we needed to test regarding the chest cavity of our dinosaur. Alkaline stomach contents might have affected the skin preservation in the chest cavity, but where stomach contents did not play a role, humic acids might have been able to release specific inorganic molecules to react with iron in the soil, forming siderite. That process was enhanced by the presence of flesh, yielding the carbon dioxide that gave rise to the calcium-carbonate cements. Still, an important question remained: Where did all the iron come from?

Although Dakota did not suffer attacks from too many scavengers, he did not escape entirely unscathed. During the preparation of the body block, a bony hand and arm was slowly revealed. The arm did not belong to a dinosaur, but to a small alligator-like crocodilian from the Hell Creek Formation, called *Borealosuchus*. What was it doing stuck up against, or even possibly inside, Dakota?

This was a real problem to solve. Did the crocodile have anything to do with Dakota's death? That scenario was unlikely, given *Borealosuchus* was much smaller than *Edmontosaurus*. Could the carcass of an already dead crocodile be washed up against the body of Dakota in the same event that buried Dakota? That is possible, but the bones of the croc were articulated, suggesting it had not traveled far. The *Borealosuchus* might have been scavenging the partially exposed body of Dakota. The croc might have forced its way into the chest cavity of Dakota but become stuck. It is possible that, unable to escape the rib bars of its cage, the crocodile drowned, inside Dakota.

The combination of scavengers and Dakota's stomach chemistry might be the reason for the lack of skin impressions around his chest. The scavengers might have stripped any flesh exposed to the surface, and the fermenting stomach contents of Dakota might have neutralized the all-important acid groundwaters, preventing specific molecules from being released from the encasing sediments, which were crucial to the formation of siderite.

Despite the illuminating discoveries that were being made, the preparation was going far slower than anyone could have planned. The difficulty was purely a function of the complex changes in matrix hardness and the need for delicate preservation of the skin envelope in many places. Given the twists and turns already posed by our gigantic puzzle, Tyler and I decided that the prep had to stop. To ensure we did not damage this unique find, we needed a 3-D preparation plan for the body block, now complicated by having a crocodile inserted into the equation. Once we could see the parameters, we could proceed with more confidence. Only a huge CT scanner could provide the team with such a roadmap. What we didn't know yet was how much else, inside the dinosaur, the massive CT operation would discover.

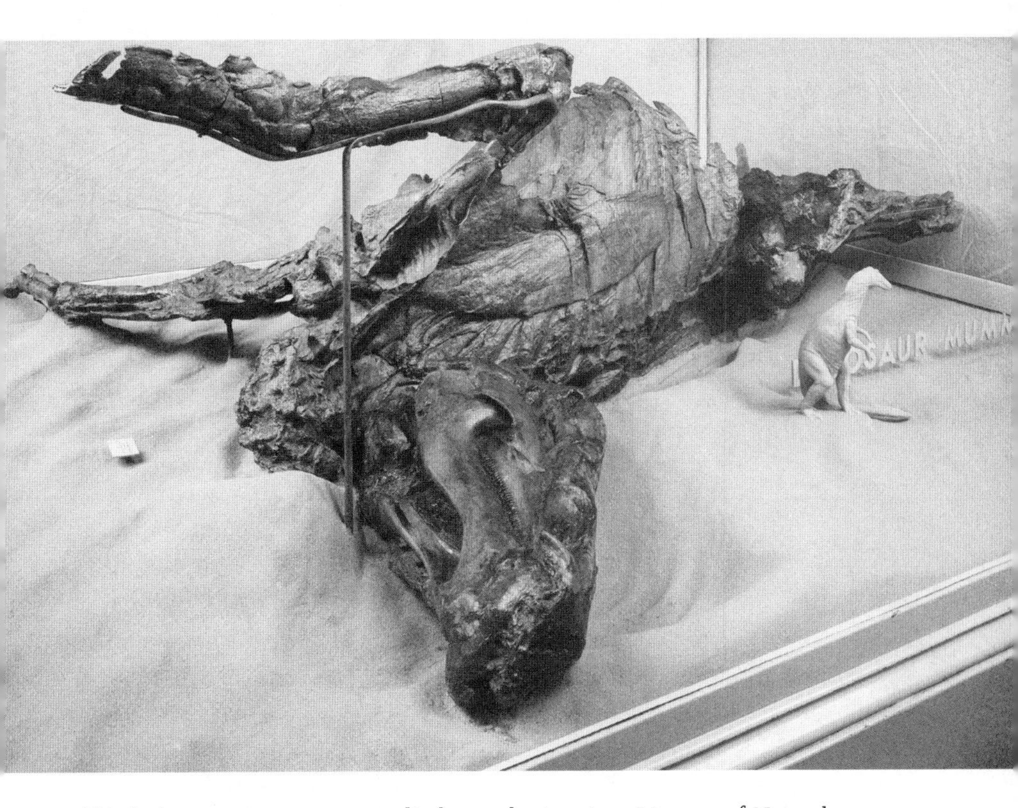

This hadrosaur mummy, now on display at the American Museum of Natural History in New York City, was discovered by Charles H. Sternberg in 1908 in the Lance Creek Formation of Converse County, Wyoming.

ABOVE: *Mineralized embryo skin of a titanosaurid sauropod dinosaur.*
RIGHT: *Excavating a triceratops in the Hell Creek Formation in 2006, Stephen Begin (left) and Steven Cohen (right) cover a vertebra in aluminum foil and tape, prior to the application of plaster to form a protective field jacket.*
FOLLOWING PAGES: *Tyler Lyson (left) and Chris Ott (right) survey the Hell Creek Formation Badlands of North Dakota.*

To get unweathered sedimentary samples, which show environmental changes over time, Dan Pepe, Chris Ott, Joe Macquaker, and Phil Manning took turns trenching a 60-foot-high butte next to the excavation site.

After exhuming the dinosaur mummy, Tyler Lyson (left) and Phil Manning (right) apply plaster-soaked burlap to the underside of the field jacket to protect the body block from damage.

ABOVE: Steven Cohen keeps a watchful eye as the precious body block is placed on the back of the flat-loader, marking the beginning of a long journey for the fossil remains of Dakota. **LEFT:** The plaster-encased body block is welded to a steel frame to give it added stability and support during transport.

A strange convoy of vehicles carefully picks its way through the Hell Creek Formation Badlands. Leading the way, a rather rare and important cargo is destined for the paleontology prep lab.

At the end of a grueling excavation, a relieved-looking field crew—from left to right: National Geographic *cameraman David Linstrom,* National Geographic *producer Jenny Kubo, Joe Macquaker, Phil Manning, and Tyler Lyson—arrives at Dakota's first stop, the Marmarth Research Foundation.*

LEFT: *The LiDAR team—Dave Hodgetts (left), Rob Gawthorpe (center), and Franklin Rarity (right)—sets up the scan station to survey the excavation site around Dakota.* **ABOVE:** *This digital outcrop model was generated from the LiDAR survey. Mapped in only five days with the help of LiDAR, an area this large would have taken weeks or months using traditional surveying techniques.*

Jeff Anders "plumbing" the up-rated 9-MeV linear accelerator into the CT housing at the Boeing/NASA facility. Positioning it correctly was critical for the generated x-rays to penetrate the enormous body block.

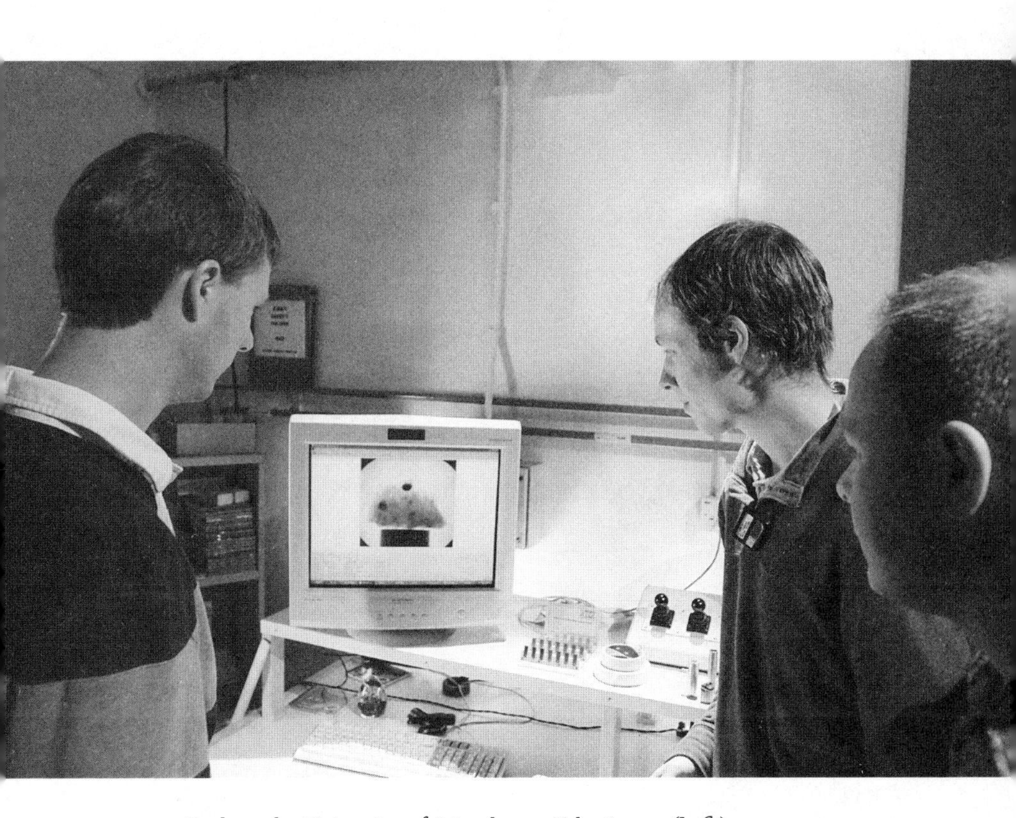

Back at the University of Manchester, Tyler Lyson (left), Sam McDonald (center), and Phil Manning (right) get a glimpse of the dense, iron-rich hoof of Dakota for the very first time. **FOLLOWING PAGE:** Tyler begins the tedious work of picking matrix, grain by grain, from the delicate surface of Dakota's fossilized skin.

CHAPTER NINE
SPACE SHUTTLES, AIRPLANES, AND DINOSAURS

"All science is either physics or stamp collecting."
—Ernest Rutherford

MEDICAL IMAGING TECHNOLOGY today enables the noninvasive imaging of serial sections through patients. Most of us are familiar with medical x-ray machines and the resultant x-ray film that allows broken or fractured bones to be imaged. The x-ray film image is a function of photons from the x-ray radiation source passing through the body of a patient showing the variation in tissue density (bone, tendon, muscle, etc.) as the photons pass through the body. The denser the material, the brighter the image, and vice versa. The x-ray film provides a cumulative density map of the whole part of the body that is examined; it does not allow you to "extract" a specific plane within that volume of flesh and bone. However, CAT (computerized axial tomography) allows the recovery of a slice of matter from within a 3-D object. The term CAT scanner is often abbreviated to CT scanner, and that will be used in this chapter.

The CT scanner works on similar principles to the simple medical x-ray film method, although it allows the reconstruction of many individual slices of matter into complex 3-D forms.

This relatively new science has its roots in the late 1950s, applied first to medicine in the early 1970s and expanded to industrial applications in the 1980s. The development of CT technology is inextricably linked to advances in computing power and storage, as the data sets generated by even small CT scans are large and complex to process. There are many types of CT scanner in terms of size, power, and resolution. The size of a scan is often tied to the resolution it can achieve, and larger objects usually equate to lower resolution. The smaller fragments of our mummy that we wanted to scan were well suited to the unit we were currently using in the School of Materials Science at the University of Manchester. However, to scan the tail and body blocks—now a mere 1,500 and 8,000 pounds, respectively, after being partly prepared—would require a much larger scanner. The body block measured 3x2 meters and the tail 2.5x1.5 meters. These would be world-record-breaking scans—if we could get them to fit into a scanner in the first place. After much hunting and with help from Dan Goldin (formerly with NASA) and National Geographic, we were able to track down a large CT facility in Canoga Park, north of Los Angeles.

The facility, located at the top of a winding road, was used not only by NASA but also by Boeing jet research, and as a result was bristling with security. We were met at the security lodge and then led in a convoy to Building 100, where the CT scanner and Jeff Anders were based. Surrounded by a ring of hills, the CT facility, rocket-engine test beds, and a strange array of industrial military facilities were hidden from prying eyes. At the time the site was coming to an end of its working life, and many of the units were being dismantled. We were soon to learn that this would be the last CT scan to take place in Building 100.

Jeff Anders came outside to greet us. Behind his bright blue eyes and calm smile is one of the sharpest minds I have had the pleasure to work with. The space program and Anders have been intimately tied together for nearly 25 years. He is one of those guys who does not want the limelight, even though he above all is one who deserves it. He walked and talked us through some of the site's impressive history. The cool morning air was beginning to warm as we headed toward Building 100. The sun was also clearing the early morning haze, reminding me that today would be filled with hot and hard work. Signs dotted the site reminding us that rattlesnakes and mountain lions were always lurking around the place. Anders told us that he had extracted several snakes from his CT facility—that year! The site was in many respects returning to nature.

To scan such a large block is not easy, especially at the resolution that we wanted, in order to see the details of skin, bone, ligaments, and possible soft-tissue structures. This Boeing/NASA unit had been scanning space program and plane parts for the past 25 years. Anders had built the unit from scratch, starting with an empty concrete shell that he entered in 1989.

As we walked inside the CT facility, the air conditioning was a welcome respite from the increasing heat. The narrow corridor from the entrance led to the CT control room. Banks of screens, computer servers, arrays of oscilloscopes, and the hum of cooling fans greeted us. In the far corner of the CT control room was a large concrete and steel door that worked on a corkscrew mechanism to open and close the barrier that sealed off the CT scanner. As we walked though the opening into the scanner cell, the 21-inch-thick steel-and-concrete door could be fully appreciated. Something told me that whatever could be generated on the other side of this door was possibly not healthy. The enclosed CT cell was gigantic, with a 50-foot-high ceiling and measuring 50x50 feet on the floor. Heavy lifting gear hung from above, capable of lifting 5,000 pounds. That would be fine for the tail block, but not for the body. The two paired steel towers of the scanner rose from the floor, ten feet high each, one pair holding the x-ray source, the opposite pair the detector array. The massive bearings beneath the turntable would have to guide our dinosaur at a right angle between the scanner source and the detector array. Running the body block through the scanner would be akin to an elephant skating though a very narrow gap while pirouetting in the midst of a million-dollar car wash. If the block so much as touched one of the towers on a scan rotation, it would prove disastrous. We had to make sure the blocks fitted.

Anders's knowledge of the CT and its breadth of application is beyond that of anyone else in the world. From space shuttle engine parts to *T. rex* skulls, nothing fazes him. Of the four awards that NASA can give to civilians, he has three. Anders was eligible for number four, but NASA did not want to set a precedent.

The sort of work that the CT unit usually undertook was finding minor imperfections or structural flaws in aerospace components, from bolts to shuttle engine parts. An example is a problem that was occurring with oxygen pumps used in the fuel system of the space shuttle. Each unit cost $650,000 to manufacture, and if one unit failed, it would have devastating consequences to the shuttle. The problem with each pump was diagnosed and solved by Anders's team, in the long term saving NASA $35 million.

In an interesting aside, he told us that dinosaur CT scans, such as Sue the *T. rex* in the 1990s, helped put astronauts into space. The physical limits of his CT machines were pushed when asked to scan an eight-foot-high block containing the skull. Anders recalled, "The very next day we used the techniques we had applied to the dinosaur to shuttle parts, as we had pushed our application knowledge with what we had achieved with the skull." Whoever said that dinosaurs are not relevant to the modern world?

The mummy blocks soon arrived from South Dakota at the CT facility. To my relief, the renewed plaster field jackets that capped the blocks had protected them on their 1,000-plus-mile journey from Hill City. However, we still had to lift them off the truck.

Neal Smith and Ruben Barroso arrived with two forklift trucks, both 15,000-pound units capable of lifting the tail and body blocks from the truck. Several tense minutes later, they had carefully removed the precious loads and delivered them to a temporary holding area outside the CT facility. Anders went straight onto the blocks as they were set on the ground by the forklifts.

"These are big!" he commented. We grabbed a tape measure and straddled the blocks to see just how big. The travel jacket protecting the blocks had increased their size, and that posed a problem. The clearance between the CT towers was 108 inches, and the body block measured 118. Anders's machine could stretch it to 114, but we still had to rotate the unwieldy block on the CT unit's turntable.

We both stood back and appraised the situation. He noticed several parts of the cradle holding the mummy projected beyond the footprint of the block. We could shave there. We agreed on other areas where the body block could be pared back. The tail block would scan already. To assist with that, I headed to the airport to pick up another member of the team, Paul Mummery, the University of Manchester's answer to Jeff Anders. Mummery had spent many years researching and developing CT scanning applications, especially micro-CT.

Mummery is one of the most cheerful academics I have ever had the pleasure of working with. He was as excited as I was about CT scanning the dinosaur. However, his enthusiasm stemmed from the perspective of a materials scientist. The thought of micro-CT scanning such huge volumes of rock was beyond the capabilities of any scanner in the world. While a paleontologist is awestruck by huge dinosaurs, Mummery gets a kick from engineering that pushes physics to the limit. Here we were on the perimeter of the outer limits.

Mummery and Anders buried themselves in their work in Building 100. Anders's calm approach revealed the years of experience that prepared him for complex problems to solve, and that would be coupled with Mummery's years of micro-CT experience. They spent several hours calibrating and checking the distance of the scanner from the detectors (all 1,024 across the array). The 6-megaelectron-volt (MeV), pulsed-source linear accelerator generating the x-rays could really spoil your day if you were stuck in front of it. The accelerator generates electrons at one end that are then contained and focused by a series of magnets, while in a vacuum. The beam of electrons has one way out, through a tungsten target at the business end of the accelerator. As the electrons hit the tungsten target they generate an intense beam of x-rays. That is what passes through the object on the CT bench. The variations in density within the material are picked up by the linear detector array on the opposite side of the accelerator.

As with most areas of science, preparation is all-important. We spent the first few days sorting mounts, methods of moving, and scan parameters. Rotating 8,000-pound blocks around on a sliding and rotating CT bench generates a lot of forces, many in the wrong sort of places. The inertia generated by the blocks' swinging would create some complex torques to deal with. After many calibrations, we placed the tail block onto the bench. Neal and Ruben helped shift the vast block into place.

By the third day at the CT unit, we finally resolved a few software glitches and the tail block was centered. By 11 p.m. the first slice of dinosaur tail appeared on the computer screen. Even in its raw form, before rendering, it looked incredible. By 8:30 the following morning we could clearly see vertebrae, plus sections showing intervertebral disk spacing, beautifully. Each slice slowly constructed the form of the tail vertebrae. Slice by slice the internal skeletal shapes of our dinosaur were appearing on the screen.

The registration of the scans was perfect, because the internal structure of each vertebra could be seen. However, this was only the first slice of 28, each taking 25 minutes. Each slice added data to the already vast file that represented our virtual dinosaur. Plus, this was just the tail—still incredibly important, but I could only think how much the body block would tell us about our animal.

By midday we had 35 slices, but we still could not combine the data into a useful rendered image until the final scan had passed. The waiting was a very frustrating time for the whole team. We avoided looking at the screen of the CT control unit, as it was only giving partial snapshots of the block's hidden contents. Not only that, each slice would take an additional 40 minutes of rendering before we could view it.

The final scan on the tail ran at a 2-millimeter aperture on the CT scanner with a huge 6-MeV pushed through the source accelerator at 95 amps. The x-ray count was shielded from us in the concrete-bunker that was our CT room. I was beginning to understand why such a facility was located in the middle of a 3,000-acre site.

The tail slices started to appear on the monitors. Slice by slice the grayscale images of the tail block revealed more and more of our animal's anatomy. The first observation that struck us all, especially Mummery and Anders, was the variation in density between the fossil bone and the surrounding matrix. The bones were dark and the matrix was lighter—was this wrong? Brighter equated to denser, and the inverse for darker regions, on the scan. Were our bones less dense than the matrix in which they sat? The brightest material was a thin veneer surrounding the tail vertebrae. Could this be our skin envelope? Until we reconstructed the slices it would be hard to say. However, the layer of bright dense material followed a course expected for the fossil skin envelope. This was promising.

The tail (caudal vertebrae) lay in an evenly spaced sequential line, slowly decreasing in size distally. The space in between each vertebra marked where the animal's intervertebral disk once connected the articular facets of the vertebrae. The regular spacing ran the length of the tail and indicated quite a thick disk, around ten-millimeters thick. The spacing would increase the overall tail length by around 50 centimeters. Coupled with the already increased length of tail due to its soft tip, containing no vertebrae, hadrosaur tails were being sold short, by nearly a meter. This finding would have implications for body mass, center of mass, and locomotion. The deep skin envelope on the prepared side of the tail already indicated the structure was more robust than previously thought. Gone were the notions of tapering skinny tail to a pointy tip.

The processing of the tail data would take many hours, but in the meantime we had a bigger problem to solve, of the 8,000-pound variety. The body block had arrived a little on the large size. The aperture of the scanner, while the block is rotating between the source and detector, would take a maximum payload of 2.895 meters. The body block was way oversize, so we had to adapt the mount. This involved several saws, a forklift, a crowbar, sledgehammer, and lots of sweat. The body block was set in its arrival building, looking very big, heavy, and immovable.

To fix the body block to the rotating CT turntable also required the construction of a special mount, to be made from several lengths of 10-centimeter box steel. Little else would be able to support such a heavy load. As I trimmed the body block's mounts, Anders scribbled a drawing for the welders on a notepad, a simple but very effective design. The frame soon appeared from the workshops, a little rusty, but certainly strong enough for the task ahead.

After a day's work the body block's support was slimmed. We shaved several inches off every corner and resupported the base. But would it fit in the scanner? First, we needed to lift the huge concrete blocks that served as a giant door, preventing x-ray radiation from leaking outside the scanner cell. This huge opening had been designed to accommodate enormous shuttle parts, but this was the first time since 1989 that they had been opened. Each of the five giant blocks had to be fork-lifted out by Neal and Ruben—representing 60,000 pounds of radiation protection. An hour later, a vast door had appeared, and the late-afternoon sun sent shafts of light into the CT scanner cell. We were ready to see if the body block would fit.

Neal drove the forklift up to the body block's storage facility. Slowly he inched the forks beneath the precious load. A little more than 300 yards away, the CT facility was not far, but it was the longest walk I had taken all day. As we inched our way toward the facility, the hydraulics on the forklift could be heard straining under the heavy payload. Each small bump in the tarmac was magnified in motion by the load extending in front of the vehicle. The forklift edged its way around the building at a slow but steady pace. Beyond the final corner into the courtyard at the back of the CT facility, the gaping door of the CT cell welcomed Neal and his cargo. The block was gently placed onto the ground. We had to figure out how to lift the block onto the CT turntable—a table that fortunately had a 20,000-pound bearing to help support huge payloads during scanning, large enough to support the gun turret of the huge M1A2 Abrams battle tanks.

We attached the box steel frame to the CT turntable. When we spun the frame through a 360-degree rotation, its circumference seemed well within the CT scanner towers' distance. Neal gingerly picked up the body block with the forklift. Carefully guiding him toward the welded frame on which it would spin, we nervously watched as he released the payload onto the frame. The steel creaked and protested as the full weight of the mummy came to bear upon the single bearing beneath the turntable. The body block and steel frame were a perfect match. With mounting anticipation, Anders engaged the motor for the turntable. The body block began to rotate slowly.

As the block rotated between the scanner towers, we realized that it was a very tight fit. The plaster jacket that protected the fossil still slightly overhung the wooden support frame on which it sat, crucially at the corners. Each time the block rotated, they came perilously close to hitting the towers. Anders scratched his head and asked me if we could take a little bit more off each corner. We had to give our dinosaur another shave.

After many hours of "tweaking" the jacket of the mummy, we were finally ready to begin. It was now very late on Thursday night. The thick concrete door that protected us from the radiation slid into place with a pleasing thud. The sirens and red strobes kicked into action and the x-ray beam was turned on.

The x-rays penetrated the huge body block, yielding a decent bulk-density scan. That meant the block could be penetrated with the accelerator's x-ray beam. If the beam could penetrate the rock, we were en route to getting the world's largest high-resolution CT scan.

The movement on the turntable was a source of worry, though. The thicker body block was gigantic, and the speed at which the block accelerated when the unit started turning was worryingly fast, akin to a rhino jumping into a sprint from standing still. The smooth action of the turntable absorbed much of the forces, but the drive belts angrily sang under the strain. The inertia of the block was so great that it was altering the relative position of the drive belts to one another, so the computer lost its point of reference, where the block sat relative to the source and detectors. Anders knew how to solve the problem, but it involved rewriting part of the software driving the motion of the scanner.

In addition, the software had decided, for the first time ever, to rotate the scan table counterclockwise, not clockwise. This added to our headache. We had to stop everything and work out why the scanner was trying to run in reverse.

To get the turntable moving the right way, we had to realign it. This meant lifting the mummy from the turntable again. To do this, we employed the 4,000-pound lifting winch from the ceiling and a pair of very heavy hydraulic jacks. Within an hour the mummy was hanging above the steel frame, leaving the turntable free to be slid from beneath the mummy. Seeing 8,000 pounds of dinosaur hovering above the ground was a little nerve-wracking.

To solve the reverse rotation problem, we decided to check the gearbox for the turntable. By 2 a.m. Anders and I had successfully rebuilt the gearbox. As for the software, it turned out that a month earlier it had received an upgrade. After a call and some cajoling with several imperious software engineers in Chicago, a new section of code was written and inserted into the CT program.

We were ready to scan the body block once again. The x-rays came on and the scan hit no hitches. A long 30 minutes passed. As the first scan became complete, Anders and I looked at it in disbelief. The density of the block was too opaque.

There was definite density variation, but the x-rays had not been hot enough, so to speak, to give us a clear signal. We would have to crank up the x-ray output from 500 to 900 rads. To put this into context, when we get a chest x-ray in a hospital it runs at 2-6 rads. Our mummy was getting a dose roughly 150 times that.

In 25 years Anders had not seen a block as dense as this one. It was denser than a block of 16-inch steel, denser than the alloys used in the space shuttle. Even at 900 rads, the scan gave plenty of evidence of density variation but no image. We would have to crank the scan up to 1,200 rads. Then, the software specter raised its ugly head. The system refused to center the x-rays on our body block. It was now the early hours of Saturday morning and we had been working hard for more than a week, with only a few CT slices of the tail to show for our trouble. We decided to call it a day. The team knew, however, that the next day might bring the breakthrough required to make the scan happen. That day's battle was drawn, not lost, and the CT technology war raged on.

In the meantime, I returned to the University of Manchester to work with Mummery and his team, in the School of Materials, where we were imaging the dinosaur's keratinous sheath from one of the toes in their own CT unit. (Unlike the massive CT scanner in Los Angeles, this one fit in a 10x10-foot room and could manage an object no larger than a human head.) We wanted to scan the sheath, so that we could model its biomechanical properties. A related subject was the dinosaur's trabecular bone. This is a spongelike internal bone that combines low density and strength with a high surface area, filling the inner cavity of almost all bones. The other main type of bone is cortical (or compact), which is denser and harder, contributing 80 percent of the weight to a human skeleton. The mechanical functioning of trabecular bone is related to bone quality, a term that includes the bone's shape, degree of mineralization, and chemical composition. Hence, the mechanical properties of trabecular bone are related to both its microscopic and macroscopic characteristics. Both of these can be addressed by considering the properties of individual trabeculae as well as their three-dimensional organization. The data from the sheath and bone would lead to a better understanding of how dinosaur skeletal materials functioned in life.

Individual trabeculae can be examined by using nanoindentation, a technique used extensively to obtain intrinsic hardness and elastic moduli values for bone tissue. The elastic moduli is the mathematical description of an object's elastic (nonpermanent) deformation when a force is applied to it. Nanoindentation offers many advantages over mechanical tests, where individual trabeculae are isolated and tested, such as minimal sample preparation and high spatial resolution. Nanoindentation studies on trabecular bone have investigated differences with compact bone as well as properties under dry and wet conditions.

The intricate architecture of trabecular bone can be imaged using the same CT scanning techniques we were applying to the tail and body in Los Angeles. Parameters such as trabecular thickness, trabecular separation, and bone volume/total volume can be quantified. This method offers distinct advantages over standard physical sectioning techniques. These measures of architecture in three dimensions can be related to the mechanical properties of trabecular bone at the cellular level. Furthermore, the elastic response of trabecular bone can be used to understand how it performed in a living animal.

Micro-CT scans were used to generate a high-resolution 3-D model for recreating the structure of the trabecular bone of Dakota. In this case we wanted to perform a stress analysis on another animal to see what would happen. The model we chose had the advantage of being based on an the actual fossil bone structure, typically with a large variation in cell size and shape. Working with CT experts Paul Mummery and Sam McDonald, in the School of Materials at the University of Manchester, we were able to generate high-resolution 3-D models for the different bone structures.

The mechanical properties of owl bone was measured by nanoindentation (a mechanical test of hardness) and used to provide the physical properties for our 3-D virtual dinosaur bone models. We choose owl bone, since in evolutionary terms dinosaurs fall between birds (as descendants) and crocodiles (as ancestors). The resulting model of the dinosaur bone was, thus, an accurate representation of the original bone structure but with its mechanical properties provided by the bird bone. In a way, we had resurrected the structure and mechanical properties of dinosaur bone.

As Mummery and McDonald were using the micro-CT, I was dissecting eagle owls (*Bubo bubo*) and tawny owls (*Strix aluco*) that had been sent to me by mail. The parcels soon earned me the nickname "Harry Potter." The birds had been kindly donated by an owl sanctuary in Gloucestershire and were either road kill or had been roasted on power lines. The owls were a great source for materials data for the FEA models we were constructing for the dinosaur bone and soft tissues.

At last I received an e-mail from Jeff Anders: The tail block of the mummy had finished being scanned, and the body block was ready to be stuck on the CT "spit" again. It was time for me to return to the Boeing/NASA facility in Los Angeles. I wanted to be present to help workout the best way to maneuver the body block back onto the CT bench. I could also start wading through the large CT images using the facility's dedicated software. This sort of data is just too big to be e-mailed!

Back in Los Angeles, and rather jetlagged, I returned to Building 100, my home for the next two weeks. As I walked into the bunker, I was greeted by a familiar, thick wall of air conditioning; a soft red beacon was flashing at the other end of the space, indicating that the CT was up and running. "We're calibrating the CT with a chunk of space shuttle," explained Anders. On the monitor I could see a distinctive bell-shaped metal structure gently rotating with a red location-tracking laser skimming across its surface. "That's one of the main combustion engines from the space-shuttle obiter," he explained.

They were hunting for a tiny flaw in the structure that had been leaking liquid hydrogen. As he informed me that this was the only CT facility that could undertake such large-scale and high-resolution work, I felt a warm glow inside. Once again I was working with a real rocket scientist.

Anders hit the kill switch for the scanner so we could take a look at the mummy body and tail blocks that were waiting inside the CT cell. The mummy would have to wait its turn for the CT scanner, given that the shuttle part took priority over the fossil. In the meantime, I started processing the mummy tail data that Anders had been re-scanning these past two months. The scanner source had died on at least two separate occasions, he noted, leaving him to rebuild the unit each time. As we walked around the scan cell, we passed a large workbench strewn with tools, cables, capacitors, and various innards of the 6-MeV scanner source. The older 4-MeV unit had been doing its best for the last few weeks, since its more powerful predecessor had died. That would not, however, be powerful enough for the body block.

Fortunately, Anders knew where to locate a 9-MeV linear accelerator x-ray source, which was not being used by a company called HESCO in Alameda, a town in the San Francisco Bay Area. Theoretically, we could marry this more powerful source straight onto the Boeing/NASA unit. Again, only Anders would be able to pull off such a feat.

We contacted Mike Depinna and John Powis at HESCO. Anders and I explained what we were doing with the CT scanning and the problems we had been having with the 6-MeV unit. Although we had recovered some slices, the resolution was hampered by the sheer density of the blocks. A 9-MeV unit would solve this problem, slicing through the blocks like a hot knife through butter. Depinna and Powis kindly offered the unit for loan with no charge. This very generous offer meant we could start roasting the mummy once more on the CT spit. However, Anders pointed out that we would have to place a large exclusion zone around the back of Building 100, as the

9-MeV unit would kick out nearly 3,000 rads that would easily penetrate the mummy, the CT machine, and the six-foot-thick back wall of the CT cell.

Images of the 9-MeV linear accelerator arrived shortly by e-mail. This unit was more than twice as long as the existing 6-MeV unit. Plus, the business end of the accelerator was large and complex with several additional water cooler inlets and valves and the tungsten target frustratingly close. Anders was not expecting such an assembly, but it was already en route and we had to figure out how to join the unit with the existing system.

Once again I was faced with not knowing if I'd get the body block scanned. If the unit that was winging its way from HESCO would not work with the Boeing/NASA unit, once more I would be stuffed. Anders sat staring at the image of the linear accelerator on the monitor, trying to make sense of the strange unit we had been offered. If anyone could do it, I was confident that he would be able to. As Anders labored on with the images of the HESCO linear accelerator, I turned my attention back to the tail block CT data.

After several hours of learning the software, I managed to bring up the first sectional views of the tail. The scan data was still a little grainy, as the middle slices had been completed using the 4-MeV accelerator after the 6-MeV had died. Still, I could clearly see that the tail was not lying completely flat. There was an undulation to the tail on its side, with a gentle twist throwing the bony processes on top (neural spines) and below (chevrons) in and out of each scan plane. The slices allowed an image for the first time of a cross-section through the tail, beyond where it had parted from the body block. Each vertebra could be seen in clear cross-section, as could fine traces of the lattice of tendons that crisscrossed the neural spines. The side view of the tail revealed what we had already guessed, that the tail was in perfect articulation, each bone in the correct position relative to where it had been when the animal was alive. More important, the skin envelope could be distinguished. A slightly brighter, so relatively dense, envelope of "skin" surrounded much of the tail. This was excellent news. We could start calculating tail muscle volume, constrained by the skin envelope, but I needed cleaner images of the core of the tail block. Bring on that 9-MeV linear accelerator!

Anders was still working out the plumbing for the 9-MeV unit so it would fit into the 6-MeV casing. It was going to be very tight, and the end of the replacement unit would hang at least three feet out of the back of the cabinet. It was not going to look pretty, but it would be functional, we hoped. We also needed to replumb the water coolant system and replace the 15-kilovolt transformer with an 18-kilovolt unit. We would spend a few days of hard work before knowing if the effort had been worth it.

Anders and I arrived early at the CT facility the next morning. Before long the call came from Sami at the entrance security gate that HESCO had arrived. Mike Depinna and John Powis pulled up in a big white pickup with the large linear accelerator securely strapped down. The photographs they had e-mailed had not done the unit justice. The new accelerator was roughly four times longer than the deceased one, leaving a few millimeters clearance at best within the confines of the CT x-ray source housing. We took the guys from HESCO out for lunch, toasted their generosity, and then headed back to the CT facility to fit the new chunk of hardware. After looking at all the connectors, we discovered we were missing one. Luckily, Anders had identified an engineering company in North Hollywood that could, if need be, construct the missing joint between the accelerator and the circular elements of the unit.

Chuck Heinrich, a colleague of Anders, joined in the fun of deconstructing and then rebuilding the CT unit. Looking up from a tangle of pipes, metalwork, and wiring, Anders grinned: "This is like fitting a Ferrari engine into a Ford Fiesta!" Heinrich and I laughed at the image, which was not far from the truth. As we threaded the 100-pound, 9-MeV accelerator into the cabinet, we had fractions of inches to spare. We ended up laying the

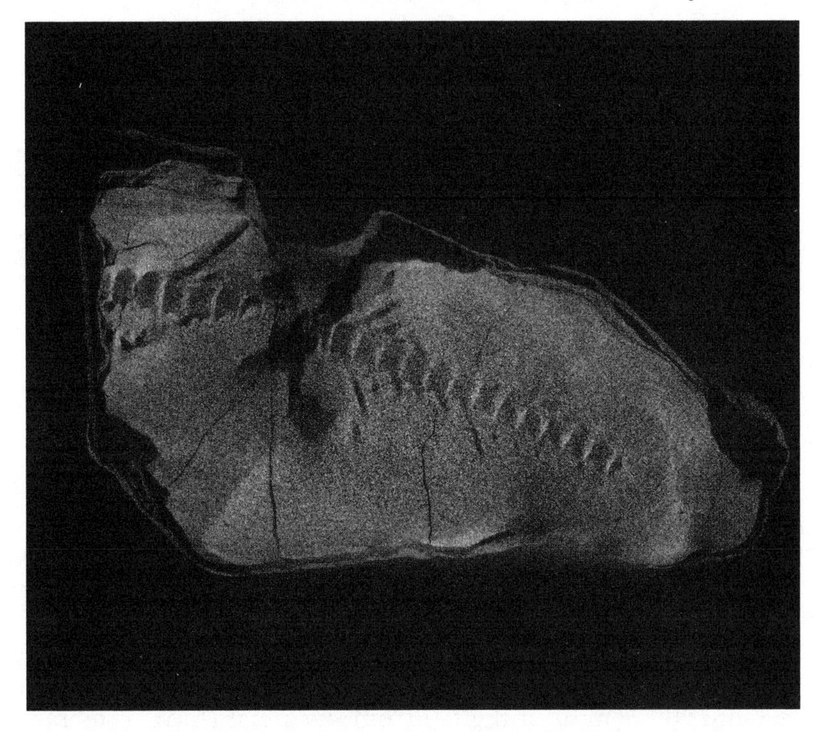

ABOVE: *This x-ray slice through Dakota's tail was generated after many hundreds of hours of preparation and scanning at the Boeing/NASA facility in Los Angeles. The intervertebral disk spacing is quite clear, indicating that soft tissue remained in place long enough to accurately preserve the relative positions of the bones.*

accelerator on its side, as it was the only way we could squeeze it into the cabinet. "That'll work," said Anders. "Just needs a bit of replumbing," commented Heinrich.

In the midst of this complicated rebuild, an e-mail came through to Anders. A decision had been made by Boeing to close the CT facility and relocate it to Florida, in four weeks' time! Blimey! We could not believe the timing. We had to work even faster. We started building a frame out of the back of the old cabinet to support the "enlarged" accelerator that was now hanging in space via temporary support straps. Several lengths of aluminum needed slicing, drilling, and taping to the side of the case. By the early hours of the morning, we had completed building the support for the accelerator and marking up the lead housing that would fit over the business end of the accelerator, as it needed drilling-out to accommodate the large 9-MeV head.

I only had two working days left at the facility. I had to be back in the United Kingdom for the weekend, because new Ph.D. students were arriving on the following Monday and I had to be there to greet them on their first day. It was looking like I would, yet again, go home without scans of the body block.

Some weeks later in mid-October, while at the Symposium of Vertebrate Paleontology (SVP) meeting in Austin, Texas, I got the call from Anders. He had just completed rescanning the tail block at a higher resolution using the 9-MeV unit. He was ready to place the body block back onto the CT spit. This was stunning news, as I was now desperate for the body block data, just to define the length of the tail from the sacral (hip) vertebrae, to combine it with the data we had from the tail block. This would help us so much in defining the size of the animal and its center of mass—so critical for the locomotion work. Not knowing whether to jump straight on a plane to Los Angeles or wait for e-mailed images to wing their way to me, I sat patiently in Austin.

The Manchester paleontology research group had five abstracts accepted for the SVP meeting in 2007. Since I was the only academic member of the staff who attended the meeting, it would have been bad form for me to leave without good reason. Two of my Ph.D. students were also at the meeting, so there was a chance I could sneak away to Los Angeles to play dino-CT. As the hours stretched into days, the telephone call from Los Angeles did not come. Many paleontologists at the meeting were aware of the CT scanning and would greet me with the words, "Any news on the CT?" After a while this question was beginning to hurt.

I learned later in the week that Anders had returned the body block to the scanner, but the CT unit had been flooded out by torrential rain. As if that was not enough, by the end of the week the weather had turned hot, dry, and windy, leading to an outburst of fires in the Los Angeles area. It seems my dinosaur had strayed into the middle of an Arnold Schwarzenegger disaster movie. As I watched the news at the end of the conference week in Austin, I could not believe my eyes.

An e-mail message from Anders confirmed the worst. The CT facility had been cut off by the flames. The e-mail had some good news mixed with a little bad. Anders started with words that I had been longing to hear, "I have all the scans we need for the tail complete. The body block is scanning now, started Friday, but lost 24 hours of scanning on Monday; no one was in to restart the scan loop so I shut down the x-rays remotely; didn't know how the fires would effect the internet connections so I shut down while I still could." After a hot, dry wind blew with gusts of 40-80 mile per hour in the valleys, reaching more then 100 miles per hour around the mountain peaks, the wind thankfully changed direction, allowing him to gain access to the CT building, which stood in the middle of the NASA/Boeing facility like a concrete monolith.

Once inside, though, he discovered further problems. The 9-MeV accelerator was not penetrating the body block. Anders used some steel blocks to figure out the true output level of the up-rated accelerator, only to find it was generating radiation like a 9-MeV, but the penetration was lower than expected. The radiation output was more like that of the old 6-MeV accelerator. The new unit was not giving enough radiation to penetrate the block. It also did not match the image quality that Anders had achieved with the 6-MeV unit, which had produced a much harder and more penetrating beam with a lower radiation count. The 9-MeV accelerator was kicking out softer x-rays that diminished within the first foot of the body block—not nearly deep enough.

To add to his problems, Anders was now receiving pressure from above to strip down and ship the smaller CT systems from the unit to a new site in Florida. The big scanner would be next. Time was running out for the mummy. Still, we had a few weeks' grace, in which time Anders could rebuild the older, harder-beamed 6-MeV accelerator. This was a man who never gave into defeat.

My last e-mail from him, before this book was due to the publisher, was not what I wanted to hear. "It's scanning again, but the bloody thing caught fire last night!" The insulation covering the 30-kilovolt electron-gun wires to the accelerator had reacted to the silicone wires, radiation, and ozone in the CT case. The whole started a nice corona, failed, and then burned! Anders also found a couple of failed high-voltage capacitors, most likely due to high-voltage arcing before the breakers tripping due to the fire. He was at least able to retrieve five scans from the body block, and these images would be processed over the next few days. As far as the pages of this book are concerned, we would be unable to report the exciting findings.

This is far from an end to the CT part of our story, but for the time being, we will have to leave our body block with some of its secrets firmly locked inside. I am sure that, as soon as the first pages of this book hit the printing press, images of our mummy's body will surface from the CT unit.

CHAPTER TEN
THE DELICATE MOLECULE OF LIFE

"The great tragedy of science—the slaying of a beautiful hypothesis by an ugly fact."
—Thomas Henry Huxley

***JURASSIC PARK* RESURRECTED** prehistoric life with a thoughtful approach toward genetically engineering dinosaurs. The key step to resurrecting a dinosaur would rely upon DNA, the molecule of life. However, the DNA blueprint for every organism is not robust but a delicate structure, susceptible to dissolution by groundwaters surrounding a carcass; it is largely gone long before the fossilization process has even started. So what is the chance that such a molecule could have survived for more than 65 million years?

To answer this question, we need to ask another first: What is DNA? It is widely understood as the chemical blueprint carried in all organisms. However, the operational details of DNA, though elegant and remarkable, are not quite so familiar. A brief review is in order, as an understanding of its chemistry is necessary in order to address questions of fossil DNA, its deterioration, the proteins which it encodes, and their bearing on my eventual discussion of dinosaur biomolecular preservation.

Every living thing known to science is based ultimately on DNA, which is present in the cells of every plant and animal, from elephant to bacterium. If DNA provides the "language" for the physical expression of this molecule, the genome is an organism's "complete works."

James Watson and Francis Crick at Cambridge University first elucidated the now famous DNA double helix in 1953. DNA was recognized as the bearer of a cell's hereditary information. Deoxyribonucleic acid is the full name abbreviated by the acronym DNA, and the term identifies this molecule as one of the two types of nucleic acid. Nucleic acids are built from chains of nucleotides that spiral around each other, staying entirely parallel. The DNA molecule's double helix looks like a twisted ladder; its backbone elements are composed of sugar-phosphates, and nucleotide pairs form the base-pair rungs of the ladder.

Surprisingly, despite the overwhelming complexity of even the simplest organism, there are only four nucleotides (adenine, thymine, guanine, and cytosine, abbreviated A, T, G, and C, respectively) that are paired to build each rung of the DNA ladder. Each base always pairs with only one other type to make such a rung. A pairs with T, and G with C. These complementary elements are called base pairs, and from these very simple elements are created the entire complexity of the genetic code. The size of the genome varies greatly from one organism to another, and a human genome contains some 3 billion base pairs, whereas some bacteria have only 600,000 DNA base pairs.

Despite the fantastic complexity of this molecular blueprint for life, it is essentially an identical process seen in every nucleus of a single given organism. Humans, for example, carry 46 chromosomes—22 from each parent, plus an X chromosome from the mother and an X or Y from the father. DNA in the nucleus of a cell is changed by the shuffling of both parents' genetic material during sexual reproduction. Each organism's genome is as unique as a fingerprint, unless it happens to be one of identical twins.

HOW DOES DNA WORK?

DNA serves as the basis of life by providing a library of instructions for making specific proteins within the cell. Proteins are large, complex molecules that enable life functions. They can serve as structural elements, building body parts such as hair and nails, or, as enzymes, they promote the many delicate processes of metabolism and digestion. In addition, proteins called antibodies protect the body from invaders by binding to hostile bacteria and signaling other parts of the immune system to destroy them. Proteins can also serve as signaling elements between cells to coordinate metabolic processes, among many other roles; they are integral to every cellular activity. Each protein is constructed from the coded information in specific stretches of DNA called genes. Humans have an estimated 20,000 to 25,000 protein-coding genes, a surprisingly low number.

Proteins consist of large organic compounds, and like their DNA partners are chain molecules, consisting of a sequence of amino acids. There are 20 standard varieties of amino acid in terrestrial biology. The amino acid sequence that constructs the protein is encoded by the DNA sequence (gene) to that specific protein. Plants can construct all their required amino acids from organic raw materials, but animals cannot synthesize them all, so they must obtain some amino acids by eating organisms that contain them. To build structures needed by the organism, ingested proteins must be broken down into their component amino acids. Proteins can be resistant to breakdown, so they have to be tackled by the body's digestive enzymes for disassembly. Many enzymes target specific proteins, such as the enzyme lactase that breaks down the familiar dairy protein lactose.

Provided a cell has all 20 amino acids to work with, it can rearrange them into specific proteins as encoded in the genes of the plant or animal. This coordination of amino-acid building blocks into the particular proteins that make up and serve a given organism requires a great deal of information. The key to this process is encoded in the nucleotide sequence of the DNA specifying which amino acids in which quantities and in which order must be assembled to make a given protein. All this information is carried in the DNA, in the sequence of simple base pairs. The DNA employs a triplet code in which sequences of three bases—AAG, for example—signal for a particular one of the 20 standard amino acids (in this case, lysine). These coding triplet sets are thus called codons. Three of the codons—UAA, UAG, and UGA—do not code for an amino acid, but instead code for "stop" to signal the end of a protein structure, terminating the translation process.

DNA is found and remains in the nucleus. Proteins are synthesized in the cytoplasm surrounding the nucleus. So a "message" must be carried from the DNA to the cytoplasm that can then be translated to successfully construct a protein molecule. The DNA is transcribed by the other kind of "messenger" nucleic acid, ribonucleic acid or RNA. If DNA can be considered the blueprint of life and the proteins the physical expression of DNA, then RNA is the crucial link between the two.

Essentially, RNA molecules are small, mobile, single-strand elements that look like half of a DNA ladder. RNA copies DNA codon sequences in the nucleus in a process called transcription, and then goes out into the cell and translates these instructions into proteins. The translation occurs at ribosomes, which are organelles that are essentially protein-assembly machines. Feed RNA into a ribosome, and the ribosome translates the RNA into a sequence of amino acids, one by one, assembled in order. When it hits a codon for "stop," out comes the protein that the RNA coded for, and the body has one more critical element needed for life processes. The same code applies to every species in the world.

All distinct protein-coding sections of DNA, the genes, have some sections that do not code directly for proteins, although it is likely that they did once. There are many portions that are non-coding sections that are not in active use by an organism; presumably this is either scrap material or baggage from the organism's evolutionary past. These sections of "scrap" DNA can include recessive genes, stretches of DNA that were once important to an organism in its evolutionary past, but are no longer used. During key developmental stages of an embryo, the presence of specific chemicals can trigger a recessive gene into coding for specific proteins and structures. The expression "as rare as hen's teeth" is based upon the reality of socketed teeth growing in the jaws of birds, courtesy of their ancestral toothy theropod dinosaur gene being activated and socketed teeth growing during the chicken's development.

PREHISTORIC DNA

The prospect of "dinosaurs brought back to life" has thrilled audiences ever since modern storytellers first devised narrative concepts that involved the great prehistoric beasts. In 1912, Arthur Conan Doyle, the author of the Sherlock Holmes mysteries, wrote a book called *The Lost World*, in which he created a remote plateau in modern South America as an extreme refugium of otherwise extinct prehistoric species, including many dinosaurs. The premise allowed human explorers to encounter dinosaurs face-to-face. Conan Doyle's descriptions portrayed active versions of the compelling creatures typically preserved in museums only as fossil skeletons. So successful was his idea that the extreme refugium concept has been reused by many succeeding writers and movie directors. Dinosaurs have rampaged among humans through dozens of novels and movies over the years, often misleading many into believing dinosaurs and humans once cohabited on Earth.

As the planet Earth shrank more and more as it was explored in the 20th century, the premise of a still undiscovered refugium became less believable. To sustain the appealing scenario of humans interacting with dinosaurs, science fiction writers were forced to devise an updated premise: the concept of time travel, as employed in Ray Bradbury's dinosaur-hunting short story "A Sound of Thunder" (1952).

Late 20th-century audiences still wanted to see dinosaurs brought back to life and interacting with humans, but by this time the genre again required a fresh premise to sustain reasonable suspension of disbelief. The latest successor to Conan Doyle is novelist Michael Crichton. Like many authors before him, Crichton borrowed from Conan Doyle the dramatic situation of humans versus dinosaurs, though Crichton went further in imitation and even copied the title *The Lost World* for one of his books. However, Crichton used a highly original narrative device for bringing dinosaurs and humans together in his book *Jurassic Park* (1990)—the use of prehistoric DNA to create clones. Crichton was inspired by contemporary research for his clever concept, which therefore seemed very convincing to popular audiences. The popularity and influence of *Jurassic Park* warrants some discussion here, which will lead us into the real science of prehistoric DNA.

Jurassic Park's premise is as follows: Mesozoic mosquitoes that fed upon dinosaur blood and were then trapped in tree sap are sought in fossil amber. When they are discovered, they are found to contain dinosaur DNA, which scientists extract from the specimens. Damaged or missing parts of dinosaur DNA are repaired by molecular biologists, who splice in sequences of frog DNA to make a complete and viable genome that then allows the cloning of a living dinosaur. A look at research on prehistoric DNA will clarify which parts of this scenario are quite close to reality and which are necessarily fictional.

THE REAL ADVENTURE IN SEARCH OF PREHISTORIC DNA

Ancient, or prehistoric, DNA has interested paleontologists ever since the idea that it might be recovered began to be seriously entertained. The improvement of molecular biological techniques such as polymerase chain reaction (PCR) and the development of laboratory equipment for sequencing DNA has greatly expanded the scope of possible research on prehistoric DNA. As a result the field is presently one being actively explored.

The breakdown of DNA in organisms dead over long periods leaves the original molecules separated into innumerable short sections. These remnant sections may number only a few hundred base pairs out of hundreds of millions in the original sequence, and the remnant sections themselves may feature damage of various kinds. To recover a fragmentary DNA sequence from one chromosome in a fossil is thus a long way from recovering an entire genome complete in every chromosome, just as the recovery of one page is not the same as having the entire works of William Shakespeare. This matter of short sequences versus a complete genome is the primary separation of reality from *Jurassic Park*.

Another division between *Jurassic Park*'s scenario and reality is the durability of DNA. Can DNA last for hundreds of millions of years? Ancient DNA studies were launched in 1984, when Berkeley researchers collected DNA from a 150-year-old museum specimen of a recently extinct zebra relative, the quagga (*Equus quagga quagga*). Before this time, DNA had been thought to deteriorate too badly for study soon after death. Researcher Svante Paabo became a major pioneer of this field. He applied DNA extraction techniques to Egyptian mummies and other ancient human remains, successfully producing sequences from bodies thousands of years old. DNA from animals dead a few thousand years, such as extinct moas (*Dinornis robustus* and *Dinornis novaezelandiae*), has allowed scientists to compare with living species in order to determine the degree to which the extinct types are related.

DNA DETERIORATION

Serious practical problems complicate the actual recovery of prehistoric DNA, even when the original molecule remains protected and in place. The molecule of life is delicate and fragile. The double-helix structure of DNA is not especially stable, as it has to be able to unzip for transcription of the genes. Even while an organism is alive, DNA is constantly being repaired of spontaneous breaks, which have been estimated to occur more than a thousand times a day in every human cell.

When an organism dies, its DNA is subject to breakdown due to the nuclease enzymes naturally occurring within the cells. The end result of this process would be the complete reduction of DNA to mononucleotides, the links from which the chain was made. At that point all the DNA information would be destroyed. However, several factors can retard nuclease activity and DNA breakdown: Cold temperatures, rapid drying, and salt are primary conserving influences. Even under optimal conditions, however, DNA continues to break down through oxidation and other chemical processes, until finally its identity is erased or so blurred that it could have originated equally from a blue whale (*Balaenoptera musculus*) or a bluebottle (*Calliphora vomitoria*) fly.

THE PROS AND CONS OF PCR

Given the short sequences likely to be recovered from fossil DNA, effective study was impractical until the invention of PCR. This technique of "amplification" creates numerous copies, even millions, from a given DNA sequence, producing sufficient material for study from even a tiny original fragment. However, PCR will blindly amplify any DNA it hits, and it is very difficult to ensure that one has targeted the actual item of interest. For example, one study intending to amplify DNA sequences from an Ice Age ground sloth found that it had produced millions of copies of DNA from modern fungus. Bacteria and fungus readily contaminate fossil DNA samples and tend to greatly outnumber the original organism's DNA remnants, rendering contamination a major factor in such studies. Amid the abundance of contaminating fungus and bacteria DNA, one is essentially looking for a molecular needle in an organic haystack.

Other esoteric methodological difficulties of this sort are par for the course in molecular biology. Results in the especially tricky area of fossil DNA analysis must be gauged by the degree to which the investigators have rigorously employed and documented the use of every possible control measure. Laboratories and equipment must be painstakingly sterilized using measures such as bleach and ultraviolet radiation, and lab workers must maintain sterile conditions via face masks and protective clothing. Even where the most strenuous efforts are taken, contamination remains a constant problem, and results claiming to identify ancient DNA must carry the burden of proof that they are not merely identifying contaminants from bacteria, fungus, or other sources.

Despite this caution, the development of PCR has enabled researchers to greatly expand the range of studies on ancient DNA. Their reports through the early 1990s kept pushing the age of DNA successfully recovered backward more deeply into time. In 1990 a sequence of chloroplast DNA was reported to have been extracted from a Miocene magnolia fossil. By 1994, amber inclusions such as a termite, a stingless bee, and a weevil—à la *Jurassic Park*—were reported as yielding fragments of DNA sequences of Oligocene and even Cretaceous age. From within a coal mine driven into the Blackhawk Formation in eastern Utah, bone fragments were recovered from which a team at Brigham Young University extracted DNA and amplified it with PCR. Their find was dated at 80 million years of age. To great excitement, the Provo team announced in 1994 that they had obtained a DNA sequence from the bone of a Cretaceous dinosaur—the achievement of the Jurassic Grail. They identified the sequence as part of the gene for mitochondrial cytochrome b, which is not found in humans, and therefore could not constitute lab worker contamination.

Nevertheless, skepticism mounted about these seemingly impossible results with the delicate molecule of life. Laboratories strained to demonstrate the critical scientific standard of reproducing the results. An inability to repeat spectacular results has dogged such studies, sowing doubt and uncertainty about their reliability. Subsequent studies on the mitochondrial cytochrome b gene sequence obtained by the Provo team showed that the sequence actually did occur in the human genome, and therefore was very likely simple contamination. Repeated attempts at perfecting the technique continued to result in cases of obvious contamination. Independent replication, when attempted, has produced negative results. Over time all the spectacularly ancient DNA findings were dismissed as incidents arising from inadvertent contamination. Pioneering ancient DNA expert Svante Paabo declared in 2001 that he considered 100,000 years as the limit for DNA survival under unremarkable conditions, with ideal preservation extending the time limit no further than one million years. More recent studies of bacteria under ideal deep-freeze conditions seem to confirm this approximate maximum range.

CONCLUSION: A MATTER OF FAITH

For conservative authorities, the door seems to be closed on hope for the recovery of dinosaur DNA, but some researchers continue to report findings of DNA sequences from extremely ancient specimens. Mary Schweitzer's extraordinary find of flexible soft tissue in a bone from *Tyrannosaurus rex* led to her team attempting DNA extraction and even going so far as implanting their extraction in frogs to see what would grow. "We're not exactly sure what we will get," Schweitzer was quoted as saying in a CNN.com article. "While it is unlikely a *T. rex* will grow in the frog, the creature may look like a bit of each. Honestly, we are not positive what will happen." Thank goodness this article was published on April Fools Day!

In 2007, Schweitzer's team reported that analysis of the *T. rex* soft tissue showed that it contained proteins capable of triggering antigenic reactions suggesting a relationship to chicken proteins. Poor popular reporting on the experiment gave the impression that the study had shown that *T. rex* proteins more closely resembled those of chickens than any other animal. However, the study actually compared the reaction of the *T. rex* material to proteins from a total of only three animals, the other two being a newt and a frog. Given these sparse alternatives, that the *T. rex* protein tested out as closer to that of a chicken seems unsurprising. However, if the results are indeed accurate, she has demonstrated the possible survival of protein molecules from extreme prehistory.

Today some researchers who have reported dinosaur DNA maintain that their findings were authentically ancient, but few others familiar with the details of the contamination issues support these claims. Contamination seems very likely but on the other hand can rarely be proven, so the matter must rest as a choice between skepticism and faith. Since DNA seems to be another of the great dinosaur soft-tissue mirages, perhaps we should be grateful to find Schweitzer's bits of protein in DNA's place. A specific protein can at least identify a specific gene that had to be present to encode for that particular material. A reverse-gene-engineering approach from paleo-protein to DNA might be the last hope that we have of "recovering" or identifying fragments of dinosaur DNA.

To resurrect a dinosaur would require an unbroken DNA sequence. In the absence of such a dinosaur genome, many have sought to bring these leviathans back to life, albeit on the silver screen. A better understanding of how we reconstruct dinosaurs and the biomaterials from which they were built is closer to being a reality than the futile hunt for DNA. It is time to meet those who rebuild dinosaurs and identify some of the essential building blocks required for this challenging work.

CHAPTER ELEVEN
RECONSTRUCTING DINOSAURS

"All art is an imitation of nature."
—Seneca

THE FOSSILIZATION OF SOFT-TISSUE impressions or structures in dinosaurs commands special interest because of the nature of these animals—their often gigantic size and the degree to which they have captivated our imagination. Dinosaurs have been classically known to the public as skeletons mounted in museums, famously lacking in the soft-tissue department, as are the overwhelming majority of fossils dug up in the field. Their true appearance in life has long been a matter of conjecture, and public interest has thus rendered very popular those artists who effectively translate paleontological reconstructions of dinosaurs into illustrations, sculptures, and motion pictures. Such media have the potential to reveal as images what are otherwise confined to the disciplined imaginations of paleontologists, who know how to read the clues to reconstruct not a mere fantasy dragon, but an image carefully constrained by and drawn from our present knowledge of science. The accomplished natural history artist Charles R. Knight was hired as a mural painter by the great natural history museums in America throughout the early 20th century for his ability to paint visualizations of the prehistoric animals that stood before the paintings only as dry bones. Pioneering moviemakers have made their own great impression on the public with dinosaur reconstructions, from the first convincing dinosaurs in Willis O'Brien's *The Lost World* (1925) and *King Kong* (1933) to those of Steven Spielberg's *Jurassic Park* (1993). We will shortly see that even the most academic dinosaur paleontologists grudgingly acknowledge a debt to such movies. At their best such popular renditions present a reflection of current paleontological thought, but the reconstructions are necessarily guesswork in many respects, with the missing pieces filled in by artists. This is the field of animation, not science.

What direct evidence do we have of soft-tissue anatomy and the dinosaurs' original appearance in life? Only rare conditions and long odds produce cases of special fossilization. Mary Schweitzer has defined exceptional preservation as "a taphonomic mode preserving soft tissues, original biomolecules or their altered fragments, original mineralogy and/or other features normally lost during diagenesis." As the earth has been scoured for dinosaurs over the last century and a half, the one in a million odds have been met here and there. Of all the known dinosaur fossils in the world, only a tiny fraction of a percentage show soft-tissue preservation, but enough dinosaurs have now come to light that a significant body of evidence exists in the form of dinosaur soft-tissue fossils. Only in recent years have we seen the most remarkable types of such fossils. For many years, skin impressions were generally the best examples, so that is a good place to start.

DINOSAUR SKIN IMPRESSIONS

Such impressions are occasionally found, though they are much rarer than bone fossils. These impressions have provided evidence of dinosaurs possessing, in some cases, a finely "pebbled" skin, with small rounded, beadlike tubercles rather than actual reptilian scales. Texture has varied from that which one might see on a football to that covering the modern Gila monster lizard. Tubercles have varied in size depending on the degree of flexibility the area of skin required, with smaller tubercles in the joints. Skin impressions are associated with specific types of dinosaurs only in a few cases, and hadrosaurs are by far the best represented.

While enough material has been found to provide a basic sense of dinosaur skin, it is also true that dinosaurs could bear ornamentation similar to that seen in modern reptiles such as horned lizards and iguanas. The ankylosaurs took this attribute to an extreme, being armored dinosaurs whose horny scutes grew to significant size, embedded in the skin over the back of the animal, forming a studded, incomplete "shield" that was often complemented by side spikes and stubby horns, attested by their substantial bony cores. Smaller studs and protuberances guarded some other dinosaurs without this degree of skin armoring, such as titanosaurs, which bore dermal plates.

In the early 1990s, spines were discovered in a sauropod quarry in Wyoming, and skin impressions connecting spines with tail vertebrae showed that these spines stood vertically along the animal's backbone, at least along the tail and possibly along the rest of the animal. This find was a striking addition to any reconstructions of the appearance of diplodocid sauropods, usually depicted with smooth, snakelike necks and tails. Hadrosaur skin impressions sometimes show tubercles patterned in rings around raised, larger dermal features resembling conical limpets. Such minor ornamentations, in the form of bosses, spines, and small ridges that would not look out of place on a magnified iguana, were probably common among many dinosaurs in ways which we can never guess without physical evidence.

Unusual fossilization conditions sometimes provide unexpected glimpses of dinosaurian skin detail. Dinosaur trackways are known from many locations around the world, but these are ordinarily foot impressions, poorly recorded as outlines by many paleontologists with no detail of any kind. Mud (fine-grained sediment, not sandy) would have to contain precisely the right degrees of wetness/dryness and stiffness/pliability to take the impression of a dinosaur foot and then preserve its detail without adhering to the animal's foot. In only a few instances, of all the thousands of tracks that have been mapped, has this special preservation been known to have occurred. One such example was discovered by Wendy Sloboda, the young daughter of a local rancher. She took an interest in the dinosaur fossils of her local area of southern Alberta, and eventually became one of the most successful field prospectors ever to work the region. One of her first discoveries was a track from along St. Mary's River, an unusual case in all respects. The sediment in which the track had originally been made had weathered away and was completely lost; however, the overlying sediments—some of which had been interpreted as fill in the hollow of the track—had become more resilient, and survived weathering. When the underlying sediments wore away, a block of the overlying strata had tumbled out of a cliff side and overturned to reveal a perfect natural cast of one of the finest dinosaur tracks ever reported. This was what Wendy Sloboda spotted in 1987.

A team associated with Philip Currie at the Royal Tyrrell Museum of Paleontology in Drumheller, Alberta, was promptly dispatched to recover this remarkable specimen. The track cast was protected with a fiberglass field jacket to ensure that its unique detail would not be damaged, and then the fossil was sliced out of its boulder by use of a large hand-held rock saw. Currie published the print, which revealed that the pebbly tubercle skin texture seen elsewhere on hadrosaurs was present even on the underside of the track. Even more interesting, the footprint showed large, flattened, fleshy pads on the hadrosaur's soles. Finer details included the bagging of skin between the toes, and even what appeared to be the imprint of the edge of the hadrosaur's toenail. The fleshy pad configurations and the shape of the living hadrosaur's foot preserved in this fossil make it a unique source of evidence. This status has won the footprint cast permanent display space at the Tyrrell Museum, where it may be seen today.

Fossil dinosaur tracks have the potential to reveal additional information on the size, gait, and speed of dinosaurs, their locomotive evolution and also to provide clues to their behavior. Furthermore, the tracks, together with the surrounding sedimentary rocks, are a record of the global Mesozoic terrestrial environments and ecosystems. When interpreted correctly, all vertebrate tracks, not just those of dinosaurs, can assist in the interpretation of past environments, behavior, and ecology, and therefore their study has wide-ranging application to themes such as biodiversity and environmental change.

The underlying assumption of many interpretations, through nearly 200 years of literature, is that what is preserved is a surface track, such as that discovered by Sloboda. The geometric data (e.g., track length and width, digit length, number of digits) on which these interpretations and trace fossil classifications are based are therefore recorded essentially as 2-D features. However, vertebrate paleoichnology—the study and naming of vertebrate tracks and traces—has concentrated on describing the trace with little or no interpretation of track formation and preservation. The way sediments behave before, during, and after a track is formed, and the subsequent processes that may further modify a track have been essentially neglected.

AUCA MAHUEVO: DINOSAUR EMBRYOS

Dinosaur eggs were first discovered by the adventurous expeditions of Roy Chapman Andrews into the Gobi Desert from 1922 to 1930. This swashbuckling paleontologist inspired the Indiana Jones character we all know and love—not an archaeologist, as he is portrayed! Among the many fossils discovered by Andrews and his team were eggs attributed to *Protoceratops*, which Pliny the

Elder had called the griffin. These eggs confirmed that dinosaurs did indeed lay eggs, a behavior that had only been postulated to that point. Jack Horner's dinosaur egg discoveries in Montana, beginning with a nest of *Maiasaura* in 1978, further developed our understanding of dinosaur behavior by showing that hatchlings stayed in the nest and received parental care for some time after hatching. The *Maiasaura* nests revealed dinosaurs as potentially good parents, like modern birds, instead of the non-parenting creatures they had generally been imagined to be, based on the behavior of non-crocodilian modern reptiles.

Andrews's eggs had contained no embryos, but the treasure house that Gobi has proved to be over the years yielded up a theropod embryo in 1993, at a site called Ukhaa Tolgod. The preservation of such delicate bones through fossilization is quite surprising, given that eggs are normally environments in which their own moistness fosters rapid decomposition in the event of death. The stench of rotten eggs signals a situation of spoilage. Since Ukhaa Tolgod other embryo fossils have come to light as well, but a site in Patagonia offered an unprecedented degree of soft-tissue preservation.

Paleontologists Luis Chiappe and Rodolfo Coria, paleontological geologist Lowell Dingus, and a team from the American Museum of Natural History were working in Patagonia in 1998 when they discovered in the Rio Colorado Formation not just dinosaur eggs but a vast area of them, a site that had been used by numerous prehistoric sauropods possibly over many seasons. Here the fossils number in the thousands, in a layer up to 16 feet deep over an area of nearly a square mile. The unrivaled site was named Auca Mahuevo from *huevo,* the Spanish word for egg. While this find set a variety of records, including the first sauropod embryos, it produced soft tissues of particular interest here. Some of the eggs contained embryos in which even the impressions of skin was exquisitely preserved, the first time this phenomenon has ever come to light. In these examples, tiny sauropod embryos are covered in finely beaded skin that is preserved in three dimensions, not flattened as skin impressions usually are in adult dinosaurs. A distinct line of larger scales runs along one embryo skin, probably relating to spinal skin features that would develop further on the adult sauropod. Only certain eggs contained embryos with fossilized skin. Local conditions involving frequent flooding might have led to the preservation of so many eggs, but another condition must have been involved in the rarer instances of the embryo skin impressions.

The importance of preparation is always paramount, but in the case of exceptional finds, it is an absolute necessity to bring the finest work to bear, lest critical data be lost. The type of sauropod represented by the Auca Mahuevo embryos was identified as specifically titanosaur, based in part on the discovery of teeth only a tenth of an inch long, which were revealed by the steady hand of preparator Marilyn Fox at Yale's Peabody Museum of Natural History. A conservative approach is vital; one does not want to discover 90 percent of the way through the job that one has been drilling away irreplaceable soft-tissue fossils while one has been proudly exposing beautiful clean bones.

Titanosaurs are known for bearing dermal plates, but the tiny embryos show no trace of the armor studs that would later have grown if these animals had lived to become adults. This suggests a development history similar to that seen in modern armored reptiles, such as alligators, in which juveniles lack the rugged features.

MARY SCHWEITZER'S PROTEIN ANALYSES

The rare conditions that preserved dinosaur embryo skin at Auca Mahuevo led paleontologist Mary Schweitzer to examine samples from this site for possible preservation of original biomolecules—proteins produced by the prehistoric animals that might survive in traces within such extremely well-preserved fossils. Schweitzer reasoned that any process that caused the mineralization of delicate embryonic remains before they decayed must have been so rapid that the process might also have preserved traces of original biomolecules. Her team worked with eggshell from the Auca Mahuevo dinosaur embryos, avoiding the testing of embryo material itself only because it is as yet too rare and precious to subject to destructive testing. Schweitzer approached the eggshell samples in several different ways, comparing the analysis with results obtained from modern eggshells.

EDX analysis, which we discussed earlier, produced elemental signature patterns within the ranges seen for modern crocodile, ostrich, and chicken eggs, which are, of course, distinctive from the sediments and minerals deposited in the Auca Mahuevo environment. This suggested that the eggshells were largely intact originals rather than substantially altered fossils, which would have shown an altered elemental composition.

For the next stage of analysis, the eggshell samples had to be finely ground. A distinct "petroliferous" odor was noted in the laboratory whenever the eggshell was ground. By contrast, the odor did not occur when the grinder was abrading either the matrix outside the eggshell or the material inside the eggshell. Though merely anecdotal, the odor adds support to the idea that these eggshells contained original organically biomineralized material or the breakdown products of these biomolecules. Although we should initially suspend belief, we all know that the fuel which drives our cars and burns in many power stations is made up of ancient biomolecules that have been preserved as oil and coal. The scale of this "organic" molecule preservation is vast.

Most strikingly, a series of enzyme-linked immunoassay analyses (ELISA), which are specialized diagnostic tests, showed that extracts from the fossil eggshell produced antigenic results similar to those obtained from modern eggshell. The reactions were not strong, but definite and distinct from background elements. Schweitzer tried several different approaches and controlled for various possible contaminants or false reactions, but obtained consistent results, suggesting that the anti-sera were responding to actual sauropod antigens—which is to say that breakdown products of original dinosaur proteins were still present in the eggshells. This does not by any means prove perfect survival of the dinosaur organic compounds, but it does show that some traces of such proteins have lasted for more than 80 million years, even if they are incomplete or somewhat altered from their precise original configuration.

Conventional wisdom in the scientific community had regarded this as impossible, but dinosaur discoveries are changing the way we understand the structural molecules of life. Schweitzer is pioneering a new mode of study in paleontology. Well aware that numerous modifications to organic molecules are likely to have occurred during such long periods of preservation, she is working to develop study methods suitable to detecting the products of chemical degradation in ways that might allow the tracing of even altered biomolecules back to their original forms. That can be accomplished through better understanding of the pathways by which the biomolecules degrade. Such work could provide startling results if tested on the fertile amounts of fossils being found in China.

SINOSAUROPTERYX AND THE CHINESE DINOSAUR FEATHERS

The fantastic window into prehistoric life provided by the fossils of Liaoning province, as discussed earlier, includes a wide variety of birds, mammals, amphibians, fish, insects, and other such creatures. Liaoning is famous primarily because of its soft-tissue preservation, especially of soft tissue that belonged to dinosaurs.

The spectacular dinosaur fossils revealed at Liaoning beginning in 1996 included forms that clearly bore evidence of having been feathered, dramatically altering our understanding of the history of feathers and the evolution of dinosaurs and birds. While the link between dinosaurs and birds had become increasingly apparent in the wake of John Ostrom's description of *Deinonychus* in 1969, it took such hard evidence to propel the idea of feathered dinosaurs to broad acceptance.

The first dinosaur discovered at Liaoning was *Sinosauropteryx*, and when published in 1996 this find shot Liaoning to worldwide fame because the fossil showed impressions of the animal's coat of hairlike proto-feathers. *Sinosauropteryx* was a small bipedal theropod, similar in skeletal form to the genus *Compsognathus*. The remains of several individuals have been found so far. The largest specimens reach three feet in length only because of their long tails. *Sinosauropteryx* is the earliest known dinosaur to bear featherlike structures—that is, if you consider *Archaeopteryx* to be a bird. The impressions show a fibrous coat of fine filaments with a very simple structure that were similar to the down covering a hatchling bird that has not yet grown its adult feathers. Rather hairlike in nature, the covering would have been akin to short filamentous fur in appearance. The proto-feathers seen in the Liaoning *Sinosauropteryx* fossils would have left no trace in ordinary fossilization, and this striking feature of the small tyrannosaurid would have remained unknown.

The proto-feathers of *Sinosauropteryx* are all the more intriguing because this dinosaur is not closely related to the famous *Archaeopteryx* of the preceding Jurassic Period, which has long held the title of "first true bird." It had traditionally been assumed that proto-feathers would be found only on close relations of *Archaeopteryx*, or on the dinosaurs related closely to the group that led to the true birds. Yet if a distant cousin like *Sinosauropteryx* also bore proto-feathers, then perhaps so did many other dinosaurs positioned evolutionarily between it and the *Archaeopteryx* line. This connection might help us locate and reconstruct many more theropod dinosaurs with this distinctly avian characteristic. That proposition would put putative feathers on dinosaurs such as *Deinonychus*, *Velociraptor*, *Oviraptor*, and other such theropods.

The small early tyrannosaurid *Dilong paradoxus* is another of Liaoning's theropod dinosaurs. This genus stretches back to the early Cretaceous Period our knowledge of the evolutionary family that produced the much larger *Tyrannosaurus rex* 60 million years later. The *Dilong* tryannosaurids, Liaoning reveals, were also covered with proto-feathers. More than a dozen feathered dinosaur genera are now known from Liaoning, with a new genus appearing in the literature nearly every year.

Although *Sinosauropteryx* and *Dilong* feature only simple proto-feathers, other birdlike small dinosaur finds from Liaoning show true pennaceous, or contoured, feathers, such as *Protarchaeopteryx* and *Caudipteryx*, although these animals appear to have used such feathers for display rather than for flight. *Microraptor*, however, described in 2003, bears feathers with the classic avian asymmetrical structure adapted for airfoil effect, suggesting that these feathers were indeed used to support flight. At a little more than two feet in average length, this dinosaur was one of the smallest of all known dinosaurs, in the range of common birds today.

DINOSAUR SOFT-TISSUE MIRAGES AND MISHAPS

Dinosaurs hold such compelling interest for so many people that any evidence with the potential to reveal new facts or shed light on old mysteries is now attended with great media fanfare. Dinosaur soft-tissue fossils, as a class, have become a kind of Holy Grail to many fossil hunters and researchers—a goal so alluring that the search has caused errors in judgment even among high-ranking, well-regarded professionals. When people want something badly enough, they often develop the ability to see what they want to see, regardless of what is actually there. Thus, along with the legitimate finds of great value and scientific interest, the history of dinosaur soft-tissue fossils is also spotted with eagerly promoted miracle finds that, on closer inspection, turn out to be mirages.

In most cases, these are not hoaxes or deliberate misrepresentations. Rather, emotional excitement at the spectacular possibilities, or a thirst for fame and recognition, has overcome coolheaded and clear-sighted analysis. This background of occasional, but prominent, mirages in dinosaur soft-tissue fossils serves as a set of cautionary tales for anyone encountering what appears to be a "find of the century."

The marvelous Liaoning shale beds produced in 1997 yet another remarkable find rendered specially interesting by its trace impressions of soft tissue forming feathers. Only in this case this fossil showed them appearing not on a primitive bird, but on a creature that bore specific skeletal traits of a dromaeosaurid dinosaur. Like so many other prize finds from China, the specimen was taken out of the country under legal conditions that the Chinese government regards as smuggling. When the feathered dinosaur showed up at a fossil and mineral trade fair in Tucson in 1999, it was purchased by dinosaur artist and paleontology writer Stephen A. Czerkas and his wife, Sylvia, for their museum in Utah. Czerkas regarded it as a vital clue in the complex mystery of the evolution of birds from dinosaurs, for here was a creature that was clearly half-birdlike (in its arms) and half-dinosaur-like (in its stiff dromaeosaurid tail and foot structures). Czerkas called it "a missing link between terrestrial dinosaurs and birds that could actually fly," and asserted that "this fossil is perhaps the best evidence since *Archaeopteryx* that birds did, in fact, evolve from certain types of carnivorous dinosaurs."

Instead of being subjected to the standard scientific peer-review process before a press announcement was made, the find was published in NATIONAL GEOGRAPHIC magazine, which named the fossil *Archaeoraptor liaoningensis* and offered a headline saying, "We can now say that birds are theropods just as confidently as we say that humans are mammals." Unfortunately, the fossil turned out to be a composite, made from fossils of at least two different animals. It was literally what it appeared to be—half primitive bird, half dinosaur. In China, paleontologist Xu Xing had come across the complete counter-slab of the dinosaurian tail, joined to the rest of its dinosaurian body, not that of a primitive bird.

NATIONAL GEOGRAPHIC had to retract its statements in print and more or less apologize, incurring unfortunate embarrassment for the venerable institution. When the Society conducted its own investigation of how the mess had arisen, Xing's dramatic "whistle-blowing" discovery was found merely to confirm what had been repeatedly suggested all along by close analysis such as high-resolution x-ray CT scanning (conducted on the fossil at the University of Texas by Professor Timothy Rowe) and top-notch hand preparation (by Canadian fossil preparator Kevin Aulenback, who had come to Utah to work on the fossil from Alberta's Tyrrell Museum of Paleontology). Rowe and Aulenback both spotted strong evidence that the so-called Archaeoraptor was actually composed of pieces from unmatched fossil slabs, simply patched together to look good. These concerns were put in the shade by the enthusiasm of the Czerkases for their spectacular find. In the rush to make a print deadline, and under the magazine's cloak of secrecy for publicity purposes, the evidence was never scrutinized as strictly as it would have been for a peer-reviewed publication. Even leading paleontologist Phil Currie had lent the prestige of his name to the Archaeoraptor project, but he was so involved with expeditions and projects all over the world that he never had time to give the article and its extraordinary claims his full professional attention. Everyone involved ended up embarrassed by the debacle. The Czerkases had to repatriate the smuggled fossil, for which they had paid $80,000, and the celebrated duo-slab is now back home in China.

Hard on the heels of Archaeoraptor's fall from grace came a new breathless announcement of an incredible dinosaur soft-tissue discovery that seemed too good to be true. On April 21, 2000, a paper in *Science* reported on a *Thescelosaurus* discovered in 1993 in South Dakota, which had been found to have its heart intact. A dinosaur heart would be of tremendous significance in the interpretation of dinosaurian biology and the understanding of the relationship between dinosaurs and birds. Would a dinosaur heart be primitive and reptilian in nature? Or more advanced, like a bird's? Or some unique design of its own? The *Thescelosaurus*—given the name "Willo"—promised to reveal the truth at last, and perhaps end as well the debate on the great mystery of dinosaurian metabolism. Willo's heart

SCIPIONYX: A FOSSIL DINOSAUR X-RAY

More certainty was discovered in most unexpected ground. The Italian peninsula holds formations of Mesozoic age, but Italian paleontologists and field workers long searched in vain for traces of dinosaurs. Prospectors for outstanding Italian Mesozoic vertebrate fossils have had to content themselves with the excellent quality of the Lower Cretaceous fishes found in southern Italy. More than two hundred years after the discovery of the first of the exquisite Pietraroja Plattenkalk fish fossils, not one dinosaur had come to light in Italy. Yet in the area of the Benevento province, conditions in the Lower Cretaceous offered one of the nearly ideal environments for fossilization: shallow, undisturbed lagoons, similar to the Jurassic shallows that produced the famous Solenhofen limestones.

Quiet, warm waters such as these, without severe wave action or strong currents, can foster the gentle deposition of fine sediments capable of preserving life remains subjected to minimal disruption. The Pietraroja formation is known for exactly the fine-grained limestones associated with this sort of environment. Study of the formation has suggested to investigators that the lagoons in the area also experienced periods of reduced oxygen levels, a factor that also supports ideal fossilization. Anoxic waters can retard bacterial decomposition rates, potentially leaving elements of soft-tissue intact before the bodies are covered and enter the taphonomic mill that is the fossil record.

A decade would pass before one particular find was recognized as a dinosaur. Discovered by an amateur, Giovanni Todesco, in a limestone formation 30 miles northwest of Naples in the 1980s, the fossil was considered by its finder as merely a bird of some kind. Not until the movie *Jurassic Park*, in 1993, did Todesco become more interested in his find. He subsequently brought it to the attention of Italian scientists. Identified as a 113-million-year-old dinosaur, the specimen vaulted from obscurity to commanding worldwide attention.

The specimen exceeded all reasonable hopes of what the Pietraroja might be capable of producing. The first detailed description of this find documented the results of years of careful and complete preparation. Paleontologists reading the paper in the March 26, 1998, issue of *Nature* might well have wondered whether the report could possibly be true, but accompanying photographs showed that the claims were not exaggerated. This new genus of carnivorous theropod dinosaur, *Scipionyx samniticus*, was preserved with not only musculature traces in several areas, but with portions of its gut clearly visible. *Scipionyx* (literally "Scipio's claw") was named in part to honor the Roman consul Publius Cornelius Scipio, and this proud military commander of antiquity might have considered the specimen's excellence worthy of his name. Although the animal's lower legs and distal tail had not been preserved, virtually the entire remainder of the skeleton was present, articulated so thoroughly as if barely touched after death. Above all was the astonishing presence of internal anatomy features. Never before had such anatomical details been described in any dinosaur fossil.

Scipionyx is known only from this single specimen, but its investigators, Cristiano Dal Sasso and Marco Signore, declared that the individual preserved was clearly a juvenile, "little more than a hatchling." The fossil is only about nine inches long. Confirming the initial impression, examination showed that many of the skeletal elements were still unfused, like the soft ends of a human baby's bones. The body proportions alone—the large rounded eye socket, and very large head-to-body ratio—were enough to signal the individual's youthful age even at first glance.

Carnivores tend to be markedly less common than herbivores in the fossil record, owing primarily to the fact that ecological laws require carnivores to be markedly less common than the animals of prey that sustain them. This makes theropod dinosaurs, such as *Tyrannosaurus rex*, rare in comparison with, say, herbivorous hadrosaurs or ceratopsian dinosaurs. Juvenile vertebrates tend to be delicate and lightly built, especially compared with adult individuals of the same species, and so relative to adults, juvenile dinosaurs are quite scarce in the fossil record. That *Scipionyx* fit both these categories, and yet also included absolutely unprecedented anatomical detail, simply seemed to defy all probability.

Small patches of muscle tissue appear in the *Scipionyx*, most prominently in two places: near the base of the tail and in the pectoral or upper chest region. The pectoral muscle mass is so well preserved that 50-power magnification reveals acicular fibers scattered throughout the area. Near the base of the tail appears a patch of a preserved muscle group. At least three different arrangements of long fibers can be discerned here, and the investigators have deemed it a portion of the *caudifemoralis longus* muscle. In addition to traces of the powerful limb retractor muscle that roots itself in the tail, muscle tissue is also visible around the animal's colon, a phenomenon that we have already seen in the case of a *Corythosaurus* mummy.

A window into the guts of a dinosaur for the first time: What did *Scipionyx* tell us? The preservation of this specimen's soft-tissue anatomy is so superb that certain primary features are surprisingly clear and do not require extended hypothetical debate or any suspension of disbelief.

The intestine and lower alimentary canal are particularly visible. "You can clearly see its surface texture, which is lumpy and shiny, almost as you would see it after dissecting a modern animal," said Michael J. Benton, a paleontologist at the University of Bristol. The intestine is folded around itself in the animal's belly area, running through the pelvis high and close to the sacrum, and leading to a termination of the colon close to the foot of the ischium. The exterior of the intestine itself bears macroscopic transverse striations visible as small, mostly parallel folds, which the investigators interpret as the muscular wall of the organ. That a dinosaur would have intestines is hardly surprising, but the soft-tissue preservation shows a position that is farther forward in the animal than is generally imagined. The position of the visible intestines corresponds more closely to what is the animal's stomach area than its belly. The length of the intestinal tract also struck researchers as surprisingly short. Long alimentary canals tend to indicate body systems with relatively low rates of nutrient absorption. The short gut of *Scipionyx* suggests a high-absorption system in this carnivore.

The gastralia of *Scipionyx* are preserved in the perfect position. Gastralia are ventral or "belly ribs," fine riblike bones that float in the integument rather than being rooted in the vertebral canal in the fashion of true ribs. Gastralia are found in a number of saurischian dinosaurs including prosauropods, but especially in theropods. Gastralia also occur in modern crocodilians and tuataras, where they offer attachment points for muscles that can help contract the chest wall. Whether the gastralia helped respiration in dinosaurs is not yet clear. Their substantial presence in large theropods such as *T. rex* suggest that they did serve some significant function. Their close association with the intestinal mass of *Scipionyx* shows that, in this genus at least, they offer support to the lower digestive organs. The gastralia also helpfully delimit the abdomen of the animal, which is commonly a matter that, despite its importance, must be left to conjecture. One reconstruction of a dinosaur can differ greatly from another depending on how much "gut" each is given, and the skeletal evidence alone can leave broad room for such interpretation. Between the gastralia line and the apparent termination of the colon close to the end of the ischium, this fossil affords an unusual degree of guidance for reconstruction of the animal's silhouette. *Scipionyx*'s soft-tissue preservation and the inclusion of its fragile little gastralia present the image of a lean body tending toward the minimum we might imagine. While this individual is only a young juvenile, such precise anatomical information is worth bearing in mind when we approach the flesh reconstruction of other dinosaurs.

Much less clear than the intestines of *Scipionyx* is a nonetheless highly interesting feature of the fossil consisting of a darkened purplish area surrounded by a reddish-brown "halo" within the body, in the area contacted by what would be elbows on a human. The reddened coloration is interpreted as hematitic, or containing iron oxide, which typically presents this type of coloration. Lying below the sternum but above the intestines, the dark body within *Scipionyx* may very likely be the fossil evidence of the animal's liver, which is more likely to leave fossil traces than other, less-dense organs, and which may well have presented a similar coloration in life. While the lungs do not show fossil traces in *Scipionyx*, the liver may also be helpful in establishing a boundary delimiting the potential position of the missing lungs, which would bear on the question of whether theropod respiration was accomplished through organs similar to those seen in modern crocodilians, or of whether theropods had lungs arranged in the much more efficient configuration seen in modern birds.

Dal Sasso and Signore also note what may be traces of tracheal rings within the muscle fibers in the area just above the sternum at the base of the neck. The extraordinary *Scipionyx* evidence takes the study of dinosaur internal anatomy out of the realm of guesswork where it has hitherto been required to operate. *Scipionyx* presents a unique view into the internal anatomy of a dinosaur, yet has no trace of integument preservation: No skin impressions or even outlines have been discerned on the slab. It seems that the preservation of soft tissues is such a delicate matter that conditions are likely to be favorable only within certain narrow confines and in certain microenvironments such as, in this case, the microenvironment within the carcass as opposed to the microenvironment surrounding its skin.

T. REX BONE MARROW AND CELLULAR PRESERVATION

How detailed can soft-tissue fossils be? Could cellular information be preserved after tens of millions of years? Fossilization commonly involves organic remains being permeated by minerals in solution over many millions of years, during which time minerals come to fill in the spaces between the original biomineralized matrix. This process, called permineralization, produces the familiar heavy, solid dinosaur bone fossil. Fresh bone, of course, is typically hollow, hard bone surrounding an inner space filled with the softer and blood-rich marrow so sought after by predators and scavengers. The idea of dinosaur marrow being fossilized in any meaningful way seemed highly unlikely.

In 2005 a team led by Mary Schweitzer published a report detailing what appeared to be dinosaur soft-tissue fossilization on a previously unsuspected scale. A *Tyrannosaurus rex* femur from 67 million years ago discovered by Jack Horner's team from the Museum of the Rockies had emerged from the famous Hell Creek Formation in 2000. The femur was a little small for a *T. rex*, though it had been heavy enough to preclude transport by helicopter whole. The team had had to break it in half, and in doing so they discovered that the bone was not solid. It was hollow, unlike most dinosaur bones, which become completely permineralized. Recognizing the unusual situation, samples of the interior were taken for analysis by Schweitzer, who was working at the Museum of the Rockies at the time.

The fragments from the hollow femur produced astonishing results upon Schweitzer's analysis. After the mineral matrix was dissolved away by treatment in a solution, what remained was a flexible, resilient, elastic material, fibrous and vascular in form. Fossils are dusty, dry stone in the experience of most of us who work in the field. Fossils that you can bend and squash like Jell-O are, to put it mildly, unconventional. In fact, no one had reported anything like this before. The remnant analyzed by Schweitzer appeared to be the remains of soft tissue from the Tyrannosaur's bone marrow.

Closer analysis did not dampen this initial impression. Microscopic study only deepened the team's belief that they were somehow peering at the fossils of Mesozoic vascular tissue from within the femur. The size and pattern of the vascular material from the *T. rex* compared closely with samples prepared from the marrow of a modern ostrich. Schweitzer turned to the ostrich because this animal is remotely related to a *T. rex*.

Electron-microscope analysis showed that even the interiors of the *T. rex* vessels appeared to be patterned with small bumps that corresponded to the endothelial nuclei in the ostrich vessels.

The cortical, or outer, bone fossil samples were found to hold parallel networks of features that closely resemble the Haversian canals known in modern mammals' bones. This much was not a surprise, as Haversian canals or their close analogues have been recognized in dinosaur bone fossils for some decades now, albeit in polished thin sections that rely on slicing up the bone. Haversian canals are associated in mammals with endothermy, or a body's producing its own internal heat, and this has placed them in the spotlight in the debates over dinosaurian metabolism. Perhaps the most ardent champion, paleontologist Robert Bakker, strongly argued in the later 20th century that the fossil evidence of Haversian canals was evidence in support of dinosaur endothermy. However, further research found that the correspondence of Haversian canals to endothermy is not so straightforward. The presence of Haversian canals in dinosaur fossils is now regarded as suggestive but not conclusive evidence of dinosaurian endothermy.

For this reason, the presence of parallel vascular Haversian-like canals in the samples of bone from the Hell Creek *T. rex* femur was not as striking as the fact that Schweitzer's team could see light through these structures. Complete demineralization left thin, transparent soft-tissue vessels floating in the solution. In structure, pattern, and size, these vessels were again very similar to structures isolated from modern ostrich bone. The transparency of these tissues allowed the research team to peer inside, where they were able to photograph small "round microstructures," dark red to brown in color. These small round structures are virtually identical in size and appearance to red blood cells appearing in similar vessels isolated from modern ostrich bones.

Demineralized cortical bone from the *T. rex* showed what appeared to correspond directly to ostrich osteocytes, complete with the fine filipodia extensions characteristic of these cells.

Schweitzer's team pursued their analysis to the point that they consistently found intact structures at the subcellular level, with nuclei clearly visible and the cells bearing the fine attenuated filopodia that characterize osteocytes. Dinosaur osteocytes, fully three-dimensional, floating in solution! Dinosaur cell nuclei! Is *anything* possible with the right fossilization conditions?

And what of fossilization still further below the cellular level? Schweitzer attempted to isolate proteins from the fossil material, and was able through antigenic analysis to obtain results that suggested that some faint traces of the original breakdown products of proteins might indeed remain. In 2007 Schweitzer's team did one better; they nailed a protein molecule from the femur of a mastodon and *T. rex*.

Schweitzer tested it against various antibodies that are known to react with collagen. Identifying collagen would indicate that it is original to *T. rex*—that the tissue contains remnants of the molecules produced by the dinosaur. After her chemical and molecular analyses of the tissue indicated that original protein fragments might be preserved, she turned to colleagues John Asara and Lewis Cantley of Harvard Medical School to see if they could confirm her suspicions by finding the amino acid used to make collagen, a fibrous protein found in bone. Bone is a composite material, consisting of both protein and mineral. When they compared the collagen sequences to a database that contains existing sequences from modern species, they found that the *T. rex* sequence had similarities to those of chickens, and that the mastodon was more closely related to mammals, including the African elephant. The protein fragments in the *T. rex* fossil appear to most closely match amino acid sequences found in collagen of present-day chickens, clearly supporting the notion that birds and theropod dinosaurs are evolutionarily related.

FROM POLE TO POO?

Dinosaur soft tissue is turning up in other extraordinary ways. In part this is because the world has never been so globally connected. Virtually nowhere outside the polar wastes is remote in the way that the Gobi Desert was to Roy Chapman Andrews in 1922. The modern market has also generated legions of collectors everywhere, and fossil hunters motivated by private and scientific buyers alike comb the far corners of the Earth for new finds.

Better transportation and a brisk market do not, however, explain every aspect of the recent developments in soft-tissue discovery. As several of the investigators have noted, part of the reason stems from being more sensitive to the possibilities. As scientific investigation has developed the ability to extract more and more information from less and less data, collectors are realizing that what once was yesterday's scrap might be tomorrow's molecular fossil lodestone.

Unconventional approaches are paying off. Dinosaur muscle tissue has even been recognized in portions of carnivorous dinosaur coprolites, which are fossilized spoor or droppings. Undigested tissue in a Late Cretacous Tyrannosaurid coprolite discovered in Alberta has been studied for theropod digestive biology. The presence of muscle and connective tissue in such a clearly undigested state in the coprolite suggests rapid movement of food through the alimentary canal, with meat kept in the body for so little time that not all of it is digested, a phenomenon observed in modern animals which feed by periodic gorging. The short gut of *Scipionyx*, discussed earlier, also supports the rapid processing and quick "deposit" of fecal material.

While much remains to be learned, dinosaur soft-tissue fossils and creative approaches to both fieldwork and laboratory analysis have opened up a picture of dinosaurs that is far more detailed and complete than early paleontologists ever imagined. For every one of us who works in the field, the consciousness of the potential loss with any find that might display exceptional preservation is thus heightened—we are more aware than ever of what we may ruin if we screw up.

THE SEARCH FOR BIOMOLECULES

What is so remarkable about Dakota is that, in addition to the preservation of skin impressions, one can see a clear, albeit slightly decayed, keratinous sheath on the tips of each toe. This "hoof" is like your fingernail, only much bigger! The hoof material was preserved in siderite; chemistry had played its part once again. Whether there are still any of the original molecules that made up the skin or hoof structures remains to be tested. This work has taken many months in the labs back at the University of Manchester and is very much ongoing. We will not get any DNA, the building blocks of life, but we have been hunting proteins. As we discussed earlier, DNA is a water-soluble molecule and breaks down rapidly after death, but many proteins such as collagen and keratin are much tougher and may survive. That would be very, very special.

On one of the trips to the BHI to catch up on the preparation, I had taken a geochemistry sample-collection kit, consisting of scalpels, surgical gloves, glass vials, autoclaved aluminum foil, tweezers, and dissection tools. I was amazed that I was not stopped by immigration and asked if I was en route to the poppy fields of Central America. To collect the freshly exposed fossil surfaces, I had to use clean implements, combined with clear storage and labeling protocols. I had been firmly schooled in this area by one of Manchester's organic geochemists, Andy Gize, before I departed.

Wogelius and Gize had been very busy since the lunchtime curry we'd shared a month earlier. They had been examining possible techniques we could apply to the collected samples. They decided to use state-of-the-art biomolecule "hunting" equipment at the Wolfson Molecular Imaging Centre at the University of Manchester. Before that stage they had some existing apparatus in the lab that they used to analyze the first batch of samples from Dakota.

The fact that this particular dinosaur was incredibly well-preserved, to the point of being able to pick out skin texture and perhaps other fine structure, interested Wogelius very much, and he asked me whether anybody was looking at the chemistry of the samples, both inorganic and organic. One of his research techniques, developed in large part by his Ph.D. student Pete Morris, involved using infra-red light beams to analyze the reactions of amino acids (the building blocks of proteins) with mineral surfaces. Most organic molecules have groupings of atoms that absorb infra-red light at a certain wavelength or set of wavelengths. By passing a beam of infra-red light through a sample or bouncing the light off a surface, he could see what wavelengths were absorbed. The absorption spectrum for a material then allowed him to identify which organic groups were present in a sample. They could even give an idea of how much of a compound is present based on the intensity of absorption.

Wogelius said, "It seems to me, Phil, if you have such excellent textural preservation, you just might be lucky and get a whiff of some of the original organic molecules remaining in your sample. There are a lot of organic analysis techniques that might prove successful. I'd especially like to see if we can get anywhere with the Sting Ray."

Unique among universities in the United Kingdom, Manchester has an infra-red spectrometer equipped with a Sting Ray detector system. Instead of having a single detector, the Sting Ray detector is really an array of nearly 4,000 miniature detectors. Not only may the infra-red absorption of a sample be measured, but the intensity of absorption at a particular wavelength may be spatially mapped for certain samples. Each detector covers an area of approximately seven square microns, and so small samples approximately a half-millimeter in size can be conveniently mapped. Wogelius thought this technology might be used to give information about delicate structures preserved within Dakota. He also had a faint hope that the infra-red might help to identify preserved biomolecules—a truly exciting prospect.

As with most analytical techniques, the Sting Ray does have limitations. Perhaps the most important is that the spectra are produced by relatively simple groups within organic molecules. Therefore, the results cannot be used to identify specific proteins or to clearly determine many complex molecules. In other words, the infra-red spectra can identify fragments of molecules that may give indications in simple cases, but in complicated systems a good scientist would always want a corroborating technique. Second, an infra-red beam is a relatively weak beam. Unlike medical X-rays, which penetrate your body easily, the Sting Ray needs either a thin sample (for transmission measurements) or a specially polished surface (for reflection measurements). This proved to be an important limitation because some parts of Dakota proved to be extremely difficult to handle.

Wogelius and Morris dedicated several weeks to analyzing comparable samples and other dinosaur fossils before they examined Dakota. I thought that the skin and claw of the dinosaur might still have intact molecules. Unlike mammals, whose nails and hair are made of an organic molecule referred to as alpha-keratin, dinosaurs have claws and skin made of a more robust organic molecule known as beta-keratin. This molecule is extremely tough, and the researchers thought it just might have survived any geochemical weathering. The feathers, beaks, and claws of birds are also composed of beta-keratin. Therefore, pigeon down was used to test the Sting Ray imaging capability for beta-keratin, and the results were perfect, with barbs easily resolved and barbules evident in some areas. Cross-sectioned bones from other dinosaur bones were examined, and in some stunning cases it was like seeing the inner sections of the bones for the first time. Wogelius and I decided that whether the examination of Dakota was a success or not, this approach would be useful in other studies.

Finally, the samples of Dakota arrived. Morris and Wogelius were excited at the prospect—until they saw the samples. The claw and skin specimens were tiny, and, even worse, they crumbled when they tried to take samples. It would be impossible to polish or thin them without contaminating them. They decided to simply make the best attempt at analyzing tiny fragments "as is." After several hours of analysis, one finding became clear. The amine groups (amino acid building blocks of proteins) that indicate beta-keratin protein were in fact present. They weren't abundant or present on every grain, but some grains were apparently coated with a dark, sticky residue that showed the presence of amine groups every time they were analyzed. This made Wogelius wonder if indeed some proteins had survived intact.

Because he has analyzed many types of samples, he knows it is easy to jump to conclusions. When he told me about his discovery, I asked, "Isn't this a Eureka moment?" he replied, "Yes, a Eureka moment, followed by a 'I better not get this wrong' moment." So Wogelius, Morris, and I decided to double-check this result. The Wolfson Medical Imaging Centre could carefully analyze the samples for specific proteins. If the amine groups were located within intact beta-keratin molecules, then the proteins could be further analyzed. On the other hand, if the proteins had broken down and the amines were part of fragments, then other organic-analysis techniques would be needed.

Wogelius suggested an additional technique. The department possessed a useful piece of equipment called an electron microprobe. It uses a focused beam of electrons to excite a sample to emit x-rays. The x-rays emitted by each element are distinctive, and the results can be used to identify and quantify the elements present in a sample. This was almost certainly the first time a dinosaur had been subjected to this type of investigation.

This technique was particularly useful for mapping the location of minerals that had been precipitated during the complex set of reactions that partially altered Dakota after burial. The results were very promising. When the electron probe operator, Dave Plant, surfaced from the depths of his lab, he flashed a broad smile. The polished thin section from the dinosaur displayed strange elemental halos through the fossil skin envelope, but not in the surrounding sandy matrix. This finding resembled the earlier ESEM and cathodoluminescence observations, not to mention the FTIR results. Was it possible we had the decayed remnants of dinosaur cells in our fossil skin cross-section?

Sometimes if organic material is squashed hard against a clay mineral for any length of time, it can be so well "imprinted" that the clay will behave optically like the organic molecule, long after any trace of the biomolecule has decayed away. The initial investigations needed to be tested with some rigorous organic chemistry. This is where Andy Gize entered the organic fray.

MALDI, MATRIX, AND MOLECULES

To analyze the dinosaur skin residues, Gize used matrix-assisted laser desorption/ionization mass spectroscopy, or MALDI for short. This technique is especially useful when analyzing biomolecules such as proteins and their breakdown products. Such molecules tend to be fragile and do not take kindly to being blasted with a laser. First the molecules are mounted in a solution, called a matrix, to protect them. The type of solution depends upon the molecules being analyzed. As the sample is put on the laser target, the solution evaporates, leaving the biomolecules in a stable crystallized matrix that can be blasted by the laser. In this way the biomolecules are protected long enough for MALDI technique to be successful.

Before the sample could be tested with the laser, Gize had to prepare the skin samples from Dakota. Samples were treated with a broad-range enzyme called trypsin. Trypsin is a digestive enzyme produced in the pancreas and secreted in the intestine to help break down proteins, but it is also used to break down proteins to more easily analyzed peptides. A peptide is a molecule defined by a specific combination of amino acids, the building blocks of proteins. These peptides can then be identified by mass spectrometry, specifically MALDI.

Gize initially applied the technique on recent biological material chosen to represent the main keratin classes: pigeon feather, human hair, and human skin. The results were very encouraging. The hardest structural keratin, pigeon feather, yielded intense peaks at high masses. The next hardest structural keratin tested, Gize's own hair, also yielded high mass fragments but a third of the pigeon feather. The softest keratin, skin off Gize's thumb, did not yield high mass fragments. In effect, the stronger the keratin, the larger the fragments he observed.

The dinosaur skin presented a major problem in that it was heavily protected by extremely hard iron carbonate (siderite). In the end Gize powdered the sample of skin mechanically. Probably because of the iron carbonate, yields were low. Nevertheless, fragments were observed in the mass spectra that could not be assigned to contamination by Gize's skin. These fragments did not have a high enough mass to be an intact protein molecule.

After much thought Gize concluded that some protein-related residue was present in Dakota's skin sample. It was not pristine dinosaur skin, but some fragments of keratins were present. Once again, we had a unique find. It looked like we had nailed some amino-acid building blocks specific to the keratin. But how could we verify such a controversial claim? It was time to see what Adam McMahon and Emrys Jones could come up with at the Wolfson Molecular Imaging Centre.

HUNTING MOLECULES IN GEL

McMahon and Jones brought all the resources of a multimillion-dollar research center to help verify what Gize and Wogelius had found. To separate the molecular weights of any organic molecules present in the sample, they used gel electrophoresis. The technique would hopefully reveal any organic compound, ideally complete proteins, from separately processed samples of Dakota's skin and claws.

The underlying principle of gel electrophoresis is to use a "detergent" to flush out any proteins present, providing molecules of different size and electrical charge. The mixture of proteins is added to the top of the gel, which then has an electric current passed through it. The proteins migrate along the gel, dragged toward either positive or negative charge, and they travel at different speeds related to the mass of the protein. The smaller proteins will travel farther than the larger ones in a given time. A standard sample of known proteins is analyzed at the same time, and a rough estimate of the mass of the proteins—say, within Dakota's sample—can be assigned.

Once Jones applied the gel, any proteins present had to be stained so that they could be visualized within the gel. This was done using the most sensitive method available, silver staining, which can detect nanograms of protein. The silver within the solution binds to specific chemical residues on the protein and darkens the portion of the gel where the protein is localized. Typically, these are in bands within the gel. Any fans of the television show CSI will be familiar with the look of these gels, although their claims of the technique's crime-solving abilities is often very optimistic!

The next stage was to breakdown the protein using an enzyme, and like Gize, Jones used trypsin. Because a protein has specific breakdown products, any fragments obtained from a given protein can be predicted, and should be the same each time. If you have a way of identifying the size of these fragments, then the original protein can be identified, much the same as a word can be identified from a collection of its letters.

Jones began to process the samples from the dinosaur, slowly heating them in an extraction solution. He then subjected the solution to a size exclusion clean-up (removing all molecules below a certain size) to isolate any proteins, something Gize had not done. The samples were mixed with the "detergent" and placed on a gel. Blank samples were prepared in the same way using the sediment matrix that was encasing but not touching the organic sample. That would ensure that no contamination was present within the original sample or introduced during the analytical procedure. They needed to be absolutely sure there were no modern contaminants, since this would give us false hope of dinosaur biomolecules.

After the gels were run, Jones applied the silver staining, much like developing a photograph. As the gel rocked back and forth within the developing solution, all eyes were on the vessel, hoping for some protein material within the gel would absorb the stain, confirming the presence of proteins.

Unfortunately, the gel remained blank; there was no evidence of protein within the samples. Since the levels of protein expected were very low, McMahon and Jones decided to remove any inessential steps, such as the size-exclusion clean-up. This time when the gels were developed, the silver appeared to be sticking to some components within the skin samples, but not the sediment blanks. This sight was very exciting. Were these the proteins they had been looking for? Alas, the position of the stained species was well below the lowest of the standard protein markers. But what might it be?

The fact that the stain had been taken up by these molecules suggests that the chemical groups responsible for binding the silver were present, but not as complete proteins. McMahon and Jones concluded that the dinosaur sample likely contained small organic fragments of proteins, too small to elucidate the structure of the protein that they were derived from, but maintaining the chemical functionality to react with the silver within the solution. In one way this was good. We could be confident that no active microbes were contaminating the system, because they would have contributed a clear protein signal. This cloud really did have a "silver" lining, since it supported Gize's and Wogelius's analyses that at least break-down products of proteins, possibly amino acids, were preserved in the tough siderite matrix of the skin and keratinous claw.

In the future, we have to pursue the geochemical route. This process entails using a complex cocktail of chemical analysis involving nasty solvents. This is where the team will be heading with the fragile skin and keratinous sheath samples in the next few months. We already are sure there is something worth hunting for, given Marshall's cathodoluminescence images, Macquaker's EDX and SEM work, my ESEM analysis, Wogelius and Morris's FTIR data, Gize's MALDI spectroscopy, and the work of McMahon and Jones all support the presence of breakdown products of organic molecules that apparently originated from Dakota's soft tissue. The fact that five different labs at two different universities independently drew similar conclusions suggest we are certainly heading in the right direction.

from forms familiar to us today. Using Larry Witmer's Extant Phylogenetic Bracket (EPB), we can constrain dinosaurs within the families of birds and crocodilians. However, the vast number and diversity of dinosaurs makes it difficult to define how close the structure in question is to either crocodilian or avian biology. Although a *Tyrannosaurus rex* skeleton is very birdlike with its huge hind limbs and tiny forelimbs, it would be ridiculous to build the animal structured like a quadrupedal crocodilian. Thanks to the discovery of relatively complete skeletons of dinosaur genera and the application of the tools of the comparative anatomist, we now know which bones go where—that is to say, which bones connect to which. But the matters of joint rotation, position of the center of mass, and posture of the dinosaur remain in many cases a matter of active debate in paleontology.

The earliest reconstructions show that even the most basic aspects of dinosaur structure were not yet clear. Reptilian attributes of the skeletons led scientists to first use modern reptiles as templates. The sprawling posture of the crocodile and the monitor lizard was reflected in the earliest reconstructions of dinosaurs, as can be seen in the Crystal Palace *Iguanodon*, *Megalosaurus*, and *Hyleosaurus* that were built in the 1850s. Pioneer paleontologist and engineer Louis Dollo, who had more complete skeletal evidence to work with in the 1870s and 1880s, could see that the *Iguanodon* had long hind limbs. He therefore adopted a modern kangaroo as an analog and reconstructed the skeletons with an upright stance. Dollo also compared the posture of *Iguanodon* with that of birds, as many photographs of his laboratory clearly show avian skeletal mounts being used for reference. However, the size and weight of the fossil bones ultimately dictated the tail-down kangaroo pose seen in so many early dinosaur reconstructions. Before the advent of plastics and glass fiber, mounting such skeleton in an avian pose verged on the impossible with such large and heavy species. The result was an *Iguanodon* constructed in an unnatural, almost upright position with knees impossibly bent and tail vertebrae contorted (sometimes having been broken) to fit the upright pose. These postures we know today to be incorrect, but it has taken much effort to view dinosaurs as themselves rather than through the prism of our experience of modern animals. Only with modern laboratory materials have we been able to resurrect lifelike poses. In most cases dinosaurs are such distinctive creatures in size and geometry that modern animals cannot be used as templates to show us how dinosaurs were built, how they stood, and how they moved. We have to work out a paradigm based on evidence gleaned from fossil bones, on principles of comparative anatomy, and, more recently, by using the immense power of high-performance computing.

Every reconstruction or illustration of a dinosaur represents the educated guess of an investigator, usually influenced by the scientific consensus of a given era. They are always subject to revision in the face of new evidence. This is why dinosaurs have changed in appearance so much over the last century and a half since they were discovered: The changing forms that we give them in reconstructions reflect the dynamic nature of paleontological science as it has developed and refined its use of available data, materials, and techniques.

Even in the cases of the few dinosaur mummies discovered, the fossils do not necessarily tell us how the bones were posed in life. Dinosaur skeleton fossils are typically squashed flat by overlying strata, and frequently crushed and warped in what is called post-depositional deformation. We rarely lift an "inflated" symmetrical skeleton straight out of the ground. If you gaze closely enough at, for instance, the femurs on a museum-mounted dinosaur, you may discover that the two bones are skewed badly and do *not* look like mirror images of each other as they should. Fossil poses may be equally distorted. The skeleton may have come to rest in an awkward or un-lifelike pose. The body might have been wrested about by its predator or by scavengers. The loose carcass might have been carried along by floodwaters and piled roughly against an obstacle or sandbar with limbs tangled. Or the dinosaur's long neck might have dried in the sun and arched backward, farther and farther as the neck tendons constricted, leaving an unnatural pose for a fossil.

The bones themselves convey a limited amount of information that helps paleontologists correctly reconstruct their articulation in life. Scars on the bones show the attachment points of some of the muscles, and these valuable clues allow us to reconstruct where musculature was strung about the bones. Not all muscles leave clear scars, though, or any scars at all. Comparative anatomy and the many consistent aspects of body structure in higher vertebrates permit a good deal of reasonable inference and interpolation with regard to musculature. For example the traces of muscle tissue in the *Scipionyx* discovery can be tentatively identified as a specific muscle known in living animals, the caudifemoralis longus. In the ordinary cases of dinosaur fossils lacking soft-tissue preservation, muscle scars can provide limited information about which muscles were largest, or their relative magnitude, but they never provide a complete blueprint for dinosaur flesh. As a result they have not generally provided sufficient guidance to resolve contentious issues of dinosaur posture.

Articulation points, where bones fit together, are usually clear enough in the case of bones such as vertebrae. However, the key points of articulation that determine overall posture are often poorly preserved because key soft tissues are absent from the joints, such as cartilaginous joint capsules. Dinosaur joint articulations are quite imprecise compared to those of modern running mammals, which we can dissect to find the answer. These fossil bones are thus like puzzle pieces worn down so badly that they can fit together any number of ways; we must resort to contextual evidence to determine the correct configuration.

Guesses regarding posture have been refined over time. Sometimes posture issues are revised on a general basis. For example, we took account of the lack of tail marks in sauropod trackways and recognized that their long, slender tails must have been carried up in the air rather than dragged on the ground, as we had long presumed. Most often, however, understanding of posture is improved on an individual, species-by-species basis. The hard work of studying three-dimensional relationships proceeds from firsthand analysis of fossil bones or casts. It is virtually impossible to definitively reconstruct a tricky posture issue on the basis of reading a publication, even when illustrations or photographs are excellent. Skeletal topography is so subtle and complex that few paleontologists tackling posture questions would settle for less than either the original fossil or an accurate three-dimensional representation as the basis of their investigations. As this intensive case-by-case work proceeds, papers continue to appear in the professional literature revising our understanding of limb positions or the range of motion possible for a digit or a manus (hand) or a pes (foot), and presenting revised hypotheses and new analyses.

There are few dinosaur paleontologists in the first place, and even fewer have experience in mounting gigantic skeletons. Yet this work has historically been the best opportunity for examining the evidence regarding bone articulation, and for testing various hypotheses of how the animal might have stood in life. Rearranging skeletons and limbs to "sketch" out various possibilities can be difficult when the weights of the bones involved run into hundreds of pounds. Such practical issues have contributed to how slowly our understanding of dinosaur posture has developed. We simply have not had the tools to manipulate the evidence readily enough to support rapid progress and refinement of our ideas. Thus this rearranging of parts remains an active area of work, and new studies continue to revise our understanding of dinosaur postures. A question that has generated considerable discussion—such as the exact way in which ceratopsids such as *Triceratops* held their forelimbs—may turn out to hinge on an incorrect previous model, based on possibly a hundred years of incorrect articulation of the skeleton, due to the characteristically poor evidence of articulation provided by the fossil bones, or absence of complete elements, such as the rib cage.

Today increasingly sophisticated methods of analysis are being applied to the study of dinosaur posture. Multivariate and bivariate morphometrics—mapping the changes in an organism's shape according to its function—and the biomechanics of beam theory—measuring elasticity in relation to weight carried—have been productively applied to the analysis of the skeleton of the hadrosaur *Maiasaura*. Such tests have led to the interesting conclusion that this dinosaur probably walked primarily as a biped when it was young and small, and then dropped to become increasingly quadrupedal as it grew to adult size and weight. The growth patterns of the limb bones of individuals of different ages reveal differential strengthening over time, implying differing primary forces requiring resistance to breakage in the bone cross-sections. All of these factors are accounted for by the scenario of gradually developing quadrupedalism—a hypothesis made possible only by the helpful range of individual ages seen in collected specimens of *Maiasaura*.

DINOSAUR MOVEMENT

Movement can be properly investigated only once posture is established, but this study requires special tools. Through most of the 20th century, the only people who possessed tools allowing them to readily explore dinosaur movement were moviemakers using special-effects models. Miniature dinosaur skeletons and their joints were approximated with metal armatures and ball-and-socket joints, then covered in approximated flesh made of foam rubber, and finally given latex "skin." Such models allowed the pioneering Hollywood artist Willis H. O'Brien to represent dinosaur movement "realistically" for the first time in his 1925 movie adapting Arthur Conan Doyle's novel *The Lost World*. O'Brien was a master special-effects craftsman who achieved convincing results in his imagery on the basis of close observation of reality. The movement he gave to his dinosaur models reflected his studies of living animals, and the results on the silver screen convinced many people at the time that the footage must have been obtained in some distant jungle where dinosaurs were still alive. O'Brien's blockbuster *King Kong* in 1933 presented an array of dinosaurs and prehistoric animals that looked like the paintings of Charles R. Knight brought to life. In the dinosaur sequences O'Brien demonstrated movement possibilities for dinosaurs that seemed likely to him.

Experts such as Henry Fairfield Osborn often closely directed Charles R. Knight's reconstructions, enabling Knight's work to accurately reflect these paleontologists' theories regarding posture. Knight himself sometimes added aspects drawn from his own understanding of animal biology and behavior, such as his 1897 painting of *Laelaps* leaping into combat with a vigor that would not be accepted in the scientific community until the 1970s. Knight's vision of a sauropod standing on its hind legs to forage was a bold proposition that likewise was not approved by mainstream paleontologists until the final decades of the 20th century. In these matters of posture and movement Knight was ahead of academic paleontologists, possibly in part because he drew his inspiration from living animals rather than dry bones. Paleontologists were typically trained in aspects of zoology through laboratory dissection and work with stripped skeletons rather than through field observation of the way that animal bodies moved in the wild.

O'Brien's dinosaur models were crude approximations of real dinosaur skeletons, and they did not precisely model actual constraints on joints or range of motion. The armatures served simply to allow the artist to interpret his vision of realistic dinosaur movement in three dimensions. Armatures might have been constructed representing skeletons, with interchangeable joints built to model movements regarding articulation and range of motion, and animated to test various configurations and hypotheses regarding "walk cycles" and other movements. However, the technique of stop-motion animation was a highly esoteric art, requiring tremendous patience and a particular kind of talent. The technique has been practiced by only a few special-effects artists. Animators who could achieve a semblance of truly natural motion have never numbered more than a handful. No professional paleontologist ever took up the art of stop-motion animation in order to investigate models of dinosaur locomotion, probably because the technique was too difficult to make an experiment of uncertain value seem worth undertaking. However, in time computer technology would place the tools of such investigations of motion into the hands of anyone who could use a keyboard.

Computer modeling developed rapidly during the 1980s, and computer graphics sequences based on digital models were transformed from plastic-looking cartoons to photo-realistic images in just over ten years. By 1993 the art of computer-generated imagery (CGI) had advanced to such a degree that Steven Spielberg could astonish the world with the special-effects dinosaurs in *Jurassic Park*. This groundbreaking movie based its key digital animation sequences on stop-motion animation created first by using analog armature models, an approach that infused the new medium with the realism carried by the stop-motion tradition descended from Willis O'Brien. *Jurassic Park* featured CGI dinosaurs in motion that looked completely realistic to most audiences. This led to the extensive use of CGI for the reconstruction of dinosaurs in subsequent projects such as the BBC's *Walking with Dinosaurs* in 1999.

The motion demonstrated by these digital dinosaurs frequently looked lifelike, but it was created by an artistic eye rather than by scientific calculation. The computerized armatures underlying these dinosaurs were in fact no more rigorous than the metal ones used in the original *King Kong*. A CGI dinosaur leaping around on screen and interacting with its environment, courtesy of the wizardry of blue-screen technology, looks very believable. Our brains process the movement of the CGI dinosaur and accept its hops, skips, and jumps, but that's because we have no suitable modern analog for comparison. If we saw a CGI human hop, skip, and jump across the screen, our brain would instantly pick up on the subtle nuances in movement that were wrong, because we are so familiar with locomotion in our own species. However, due to an increasing range of software, digital armatures can be manipulated much more easily via computer interfaces than metal ones can be manipulated by hand, and paleontologists have been able to develop CG models of dinosaurs for investigating locomotion patterns.

DINOSAUR LOCOMOTION

The main dinosaur movement of interest to paleontologists has been locomotion, primarily running ability. The methods available to biologists when studying modern species, such as cinematography, video recording, measurements of energy consumption, force and pressure plate studies, are not easily applied to the extinct species—unless time travel is miraculously invented! One avenue of research has focused on sorting out the relative running abilities of the different groups of dinosaurs. Another has focused on the challenge of determining, if possible, the top running speeds of various dinosaur genera. The process of natural selection might have favored adaptations that gave rise to increased running ability and maximum running speeds. Whether the predator was trying to catch or the prey trying to escape, speed might be the key factor defining survival for an individual—or, in the long term, for the species.

Investigations of relative running ability can be approached through osteological analysis of skeletal adaptations for running, or cursoriality. This approach suits dinosaurs particularly well, since skeletal evidence is what fossils tend to provide best. Soft-tissue aspects of dinosaurs such as metabolism and muscle structure bear heavily upon the animals' potential cursoriality, but to take these factors into account requires a great deal of guesswork and speculation in most cases. In comparison, osteological analysis has the advantage of dealing with hard evidence and clearly quantifiable data. Dinosaurs, like many vertebrates, would have alternately walked and run. The relative position of the body to limbs, center of mass, and speed combines to define gait: walking, jogging, trotting, running, and galloping. Each mode of locomotion has its own specific gait pattern. Professor McNeill Alexander at the University of Leeds defines gait as "a pattern of locomotion characteristic of a limited range of speeds, described by quantities of which one or more change discontinuously at transitions to other gaits." Animals adjust their gaits to minimize energy expenditure, whether walking, hopping, or running.

Forces delivered through muscles, which exert force on a skeleton to move the limbs and body, would have driven the movement of a dinosaur from point A to B. The muscle groups themselves vary in form and function, depending on where they are deployed in the body. In living species the leg muscles tend to be more robust and stronger than the slender muscles from, say, the arm. Each muscle group is constructed of muscle fibers that are extremely long and specialized cells. Unlike most cells in the body, muscle fibers have several nuclei (where DNA is housed) spaced along the length of the muscle fiber cell, allowing better control over cell metabolism. The fibers are arranged into bundles that allow muscles to be much longer than a single fiber cell. Each muscle fiber is able, when provided with the correct nerve impulse, to develop tension and shorten itself. The shortening of the muscle fibers, which are usually attached to bone by tendons, pulls the bone to which they are attached and so moves joints in the skeleton. Muscles create movement and, with enough effort, locomotion.

The range of potential shortening, fiber length, and muscle size all contribute to how a musculoskeletal system functions, as does the rate at which the muscle can shorten. As a general rule, the faster the muscle contracts, the less force it can deliver. Muscle groups in different parts of the body will contract at varying speeds from one another, depending on the role of that particular muscle. The energy that drives the muscle movement comes, of course, from the food that an animal consumes. The economics of balancing food intake and muscle power output is critical to all organisms.

Using too much energy to run a system will result in inefficient muscle output or starvation. It is advantageous for all animals to keep energy costs low and locomotion efficient, as it is the single most expensive metabolic consumption of an organism.

Although cold-blooded reptiles and warm-blooded mammals are different in many ways, and both types may differ from the unknown internal heat generated by dinosaurs, the same physical laws apply to all terrestrial vertebrates past or present, extinct or extant. These laws, operating within vertebrate body structures of calcium phosphate-based bone and carbon-based muscle tissue, dictate certain basic parameters that we can be certain applied to dinosaurs no matter what the soft-tissue factors were. Applying these physical laws rigorously prevents building "houses of cards" in which speculation is piled upon speculation, and allows a careful scientist to draw certain conclusions that do not rely upon guesswork.

THE BELL CURVE OF SPEED VERSUS BODY SIZE

In 1978 paleontologist Walter P. Coombs, Jr., elucidated many of the basic parameters respecting dinosaur locomotion. Terrestrial vertebrate physics create an "optimal body size" for attaining high running speed, and Coombs's plot of the known (or approximately known) top speeds of living animal species against the logarithm of their body mass produces points contained within a rough bell curve. The highest speeds—the peak of the bell curve—are attained by animals weighing roughly 50 kilograms, such as antelopes and gazelles. Smaller animals may move their limbs rapidly and with great agility, but their size disallows a maximally efficient limb reach, and so even the fastest rabbit cannot match the top speed of an antelope. Likewise, larger animals encounter factors such as inertia that constrain their speeds, even when they are constructed with major adaptive features that suit cursoriality. A giraffe has magnificently long legs, but its top speed is less than half the speed of the antelope. A heavy elephant is slower yet, and cannot match the giraffe's maximum velocity even in full charge. An ordinary running elephant clocks less than half of the top speed of a rabbit. This bell curve relating body weight to top speed is a helpful reference when it comes to estimating the top speeds of various kinds of dinosaurs. If these physics issues apply equally to extinct dinosaurs as to living animals, then the very large average body mass of dinosaurs must lead to the conclusion that dinosaurs in general moved relatively slowly. Theoretical top-speed estimates for dinosaurs using this formula can be extrapolated based on an estimate of the dinosaur's body mass, although the body mass cannot be determined with high confidence due to the lack of soft-tissue evidence. Dinosaurs for which soft-tissue details are known present much better subjects for this method of analysis than do those known only from skeletal remains.

STUDYING BONES TO DETERMINE RUNNING SPEEDS

Skeletal remains, nonetheless, hold many clues to movement, and Coombs conducted a substantial review of dinosaur skeletons, tallying the number and degree of adaptations to cursoriality. Diverse modern species from multiple phyla show a consistent set of skeletal adaptations for running. The presence of the same adaptations in different phyla suggests that these represent universal constraints, which probably applied just as well to dinosaurs. Coombs therefore noted and tracked among the major dinosaur groups features such as long limbs relative to the body, small forelimbs in bipeds, and joints with hinge-like construction, which improve running efficiency by restricting superfluous motion. The resulting analysis roughly divides dinosaurs into four groups: graviportal, mediportal, subcursorial, and cursorial—corresponding to poor, fair, good, and excellent grades of running adaptation. A horse (*Equus caballus*) is an example of a cursorial type (an excellent runner), while a hippopotamus (*Hippopotamus amphibius*) is an example of a mediportal type (fair runner).

Coombs's work suggests that quadrupedal dinosaurs generally registered levels of running adaptation no better than that of hippos, supporting the classic impression of sauropods as ponderous animals. Dinosaurs generally show much less precise joint articulation than mammals commonly possess today, and sauropod joints have a high degree of play in them. That supports the conclusion of graviportal nature suggested by their high body weight and general lack of cursorial skeletal adaptations. Bipedal dinosaurs rated above-average running ability according to this approach, with hadrosaurs ranking slightly better than the modern black rhino (*Diceros bicornis*). Large theropods (predatory dinosaurs), such as *Tyrannosaurus rex*, rank slightly better still, as suits predators that must be fast enough to run down their prey, but, we will discover, this might not be the case at all. On the plains of Africa today, the fastest animals, bar the cheetah (*Acinonyx jubatus*, which can only manage short bursts of speed), are the prey animals over distance. The predators have a slight advantage on initial acceleration and possibly by being more maneuverable. This is certainly the case for the lion (*Panthera leo*) and Thompson's gazelle (*Gazella thomsoni*). The lion has an advantage in the first few seconds of the chase with its powerful acceleration, but the gazelle's mid-top speed soon outstrips that of the lion. The lion has evolved a stealth burst-of-speed approach to predation, while the gazelle has adapted, by the process of natural selection, suitable escape mechanisms. The animal that does not get away does not get the opportunity to pass its genes on to the next generation, ensuring only the fittest of each species survive. The best dinosaur runners of all, not surprisingly, were judged to be the small theropods such as *Struthiomimus*, yet Coombs's work suggests that they were probably not as fast as the modern ostrich. However, recent work indicates that many extinct species could achieve significantly higher speeds than that of an ostrich, a concept we will address later.

RELATION OF HIND LIMB LENGTH TO TRACKWAYS

Coombs's skeletal analysis focused on the relative running abilities of different kinds of dinosaurs, but the question of the actual top speeds of various dinosaurs became a matter of interest soon after John Ostrom's 1969 description of *Deinonychus*. This Cretaceous predator prompted the "dinosaur renaissance" and the reappraisal of dinosaurs as possibly active creatures rather than the torpid giant lizards they had generally been considered until then. Robert Bakker approached the question of dinosaur top speeds in 1975 by observing modern large vertebrates and deriving from their running performance a formula that related top running speed to relative hind limb length (RHL). That figure is defined as the sum total of leg length divided by the cube root of the animal's body mass. The determination of the precise leg length becomes a critical matter for this formula, since small errors can give falsely high speeds. While the RHL study offers an interesting approach to the problem, its limitations make the RHL calculation one of uncertain reliability. For one example, hind limb length would falsely indicate the speed of animals with very long, slender legs, such as giraffes.

In 1976, R. McNeill Alexander developed a formula that appeared to reliably estimate dinosaur speeds. A running animal leaves footprints that are farther apart than when it is walking. Somewhat remarkably, Alexander was able to take this fact and show that the length of an animal's stride versus the elevation of its hip above the ground are all that are necessary to calculate a good estimate of the speed at which the trackway was made. Repeated tests using extant species showed that this was the case whether walking or running, and whether the animal was on firm ground or mud. The relationship applies to animals regardless of their body forms and sizes, from humans to horses and from elephants to gerbils; thus the formula likely would apply equally well to dinosaurs. However, the critical parameter of hip height was calculated from fossil track lengths. Subsequent work on tracks has clearly shown their length to be a highly variable character that depends on the ground's substrate type and conditions. This has meant that many speed calculations from fossil trackways might be significantly under- or overestimating the original speed of the track maker.

COMBINATION APPROACHES

Ideally one would combine the valuable life record of dinosaur trackways with the skeletal evidence, allowing each body of evidence to aid in the interpretation of the other; this is precisely what authors such as Gregory S. Paul and Per Christiansen have begun to do. Paul and Christiansen use newly recognized trackways identified as made by ceratopsids to reconstruct the contentious configuration of ceratopsid forelimbs. They arrive by this technique at a moderate reconstruction between the two traditional extremes of a fully upright, elephant-like posture and a sprawling, lizard-like posture. Paul argues that this posture supports a fairly high running speed for even the largest ceratopsids such as *Torosaurus*, comparable to that of modern rhinoceros despite the larger body size.

The rhino has been particularly appealing in the case of *Triceratops*, given their common attributes of heavy, barrel-shaped bodies, herbivory, and long sharp horns. However, ongoing work by Kent A. Stevens at the University of Oregon reveals that the running-like-a-rhino aspect of this analogy is flawed. Using a three-dimensional virtual *Triceratops* model based on painstaking digital sculpting of all the individual bones of a recently prepared and excellent specimen, Stevens could pose and exercise the limbs throughout their ranges of motion. While the forelimbs of *Triceratops* were neither fully sprawling nor fully upright, as Paul and Christiansen concluded, the forelimb nonetheless moved in a basically reptilian manner, with a limited ability to take long steps, a prerequisite for an efficient run. Its shoulder and elbow joints do not project the forefoot directly in the direction of travel, as do modern fast runners like the horse or the rhino. Instead, the hinge-like shoulder joint of *Triceratops* sent the upper arm along an oblique path (from pointing backward and to the side, to just slightly forward and downward), while the elbow bent the lower forearm from directly downwards at one extreme to diagonally across the direction of travel at the other. That is not to say that *Triceratops* was incapable of forward movement; but it moved rather like a modern Komodo dragon scaled up to massive proportions (at least as far as the forelimbs are concerned; the hind limbs are a different story).

Stevens' work clearly shows that it just was not capable of long steps with the forelimbs, and the massive limbs would have been extraordinarily difficult to cycle rapidly enough to make up for that short stride length. When *Triceratops* is studied with three-dimensional models, rather than a set of two-dimensional illustrations, the true nature of this ceratopsids reveals itself. This is what makes Stevens' work so elegant and insightful. The animal traded the ability to run fast for an ability to hold its own, like a wrestler. The great horns and massive head; the horizontal, fused, ramrod-like neck bones; and the whole posture of the forelimbs (especially when their stance was broadened so that the body was lowered) suggest non-cursorial behaviors, such as jousting and wrestling for mating rights or defending against predators; these attributes would have become all the more impressive when the whole animal bounded forward explosively. When I talked to Stevens about this, he grinned and pointed out, "If a more consistent modern analog were sought, perhaps a feisty badger, lowering its forequarters and presenting a low and powerful stance, is more in keeping with what the fossilized bones of *Triceratops* are telling us."

Dinosaur speeds remain a contentious subject, and studies often seem to reflect the personalities of the writers as much as the evidence itself, because the evidence is so lacking and so many factors must be supposed in order to arrive at most speed estimates. For example, how mobile were pectoral girdles (shoulder and forelimb) in ceratopsids? We cannot say from the skeletal evidence, because the pectoral girdle "floated" in soft tissue, flesh, and tendons, and thus gives us no definite clue of where it belongs on the reconstructed skeleton, much less the degree to which it was mobile or fairly fixed in place as the animal moved. The mobility of the pectoral girdle has a substantial effect on estimated speed possibilities of quadrupeds, and yet it is entirely hypothetical. The field continues to argue back and forth of large dinosaurs that could outrun rhinos versus large dinosaurs that were slower than modern elephants.

Tyrannosaurus rex remains the lightning rod for such debates. Interest in the question of this animal's top speed is so widespread that new papers on the topic commonly rate mainstream media coverage. Authors such as Gregory S. Paul push for dramatically high *Jurassic Park* velocities, while other authors argue for sedate movement. The latter argument, based on measuring extensor muscles (the muscles that extend limbs) in modern species, claims that all large dinosaurs would have required an impossible amount of muscle mass to reach such speeds. James Farlow has even argued that a *Tyrannosaurus rex* dared not run because it would have risked serious injury if it ever stumbled and fell to the ground.

As the use of computer modeling becomes more widespread in paleontology, and precise digitized models of dinosaur skeletons become more widely available, increases in computing power are likely to make greater and greater use of elaborately fleshed-out digital dinosaurs to investigate movement and locomotion. Eventually simulations will be able to accurately model gravity, inertia, friction, and dozens of simultaneously interacting muscles, a feat presently beyond practical capacities. Such simulations have the potential to correctly reveal a wide variety of extant species' movements and top speeds. That effort is already under way with the work of William Sellers and Kent Stevens.

HOW DID DAKOTA WALK?

The question of measuring gait in dinosaurs presented a challenging problem for the mummy team, one that Sellers, Stevens, and I were keen to crack. High-quality fossil skeletons and trackways can tell us the basic shape of an animal and its stride length, but we then need to use our knowledge of the anatomy of modern animals and Newton's laws of motion to complete the puzzle.

Comparative anatomy allows us to sometimes predict where muscles should be placed on a skeleton, since muscle attachments often leave recognizable marks and patterns of muscle location. These are remarkably consistent even in relatively distant modern forms such as crocodiles and birds. However, we then need to estimate the sizes of these muscles and of the body in general. This is where Dakota is so important, since the fossil allows us to measure many parameters directly for the first time. The preserved skin provides a 3-D envelope that constrains the size of the soft tissues overlying much of the skeleton, which are primarily muscle. This evidence has allowed us to construct a more complete and accurate musculoskeletal model than is normally possible. We are closer than ever to predicting dinosaur movements by activating reconstructed "virtual" muscles of Dakota.

Being a mummy, Dakota not only tells us about skin morphology, but also helps us understand how much muscle he had in areas where the skin envelope was intact. This is important, as it can be used to understand how Dakota might once have walked or even run. Since a skin envelope encloses muscles, tendons, ligaments, organs, and skeleton, much like your socks enclose your feet, measuring the size and shape of this skin envelope helps to reveal what an animal would have looked like and, accordingly, how big its muscles could get. The arms, legs, tail, and much of Dakota's body are bounded by such a skin envelope. Fortunately, in places where the envelope is missing around the chest, the rib cage provides guidelines to body shape. Using measurements from Dakota's skin envelope, we have been able to build a computer model of his skeleton, and add muscle groups constrained by the skin-envelope data. For the first time we can make informed calculations of how fast Dakota might have been able to run. The work is still at an early stage, but it appears clear that Dakota could run—a fact that you might not expect for such a dumpy-looking animal. However, with *T. rex* on the prowl in Hell Creek, the ability to outrun the top predator of all time would be rather helpful.

Calculating the skin envelope is only half of the battle, though. It is relatively straightforward for a modern computer to calculate movement from force, but to produce walking or running gaits we need to find a pattern of muscle activation that moves the limbs appropriately, which is a much more difficult problem. There are far too many possible combinations of muscle activation to try out all of them, so we need to use artificial-intelligence (AI) search techniques to choose the most likely combinations. Even this still requires several weeks of supercomputer time to generate stable gaits. On the plus side, by proceeding in this way we have been able to produce a completely unbiased assessment of an animal's locomotor prowess. The animal is guaranteed to be able to achieve these locomotor patterns. In addition to producing impressive-looking simulations, AI techniques can also answer key scientific questions about the difficulties of locomotion in these giant terrestrial animals.

This project has been highly interactive, collaborative, and reactive to the ongoing investigation. We really did not know what new insights about dinosaur biomechanics and biology would be revealed by studying Dakota. Hence we could not predict where the focus should lie. For instance, hadrosaurs were probably bipedal and capable of switching between two- and four-legged locomotion. How this would influence activities such as turning, getting up, sitting down, running, etc., was a fascinating prospect, since no modern analog really exists.

William Sellers at the University of Manchester and Kent Stevens played key roles in unraveling the locomotion of our hadrosaur. Surprisingly little had been written in this particular area of research, the study of ornithischian dinosaurs. Plenty has been published on the locomotion of theropods, the rock stars of the dinosaur world. The research undertaken on dinosaurs is positively skewed toward the predators. Why? Possibly because we are fascinated by how unique, efficient, ferocious, and successful they were.

To understand the locomotion of our hadrosaur, Sellers and I had to start with the theropods. The technique that he had been developing for modeling bipedal locomotion for hominids had yet to be tested on dinosaurs. To use this new method on the mummy project, we first needed to publish a paper introducing, explaining, and testing this method. To date, many methods explaining the locomotion of bipedal dinosaurs have been based upon extrapolating data from their descendants, the birds. Sellers had developed a direct technique, using musculoskeletal data derived from the fossil bones and a splendid piece of software he had written. To validate the model, we first had to apply it to extant species, in Sellers's "virtual" gait lab.

He started by attempting to answer a simple question: What was the maximum running speed of bipedal theropod dinosaurs? Chasing down prey is a vital asset of extant predators, as is the avoidance of being eaten for prey animals. Sellers and I realized that such an estimate of speed would be of interest to paleobiologists, who study dinosaurs, and at the same time would allow us to test the method we were to apply later to the hadrosaur model.

We found that the range of predicted speeds in the literature was as variable as the methods chosen. Bold authors favored high speeds while the more cautious preferred moderate or low speeds. The current techniques were based on anatomical comparisons (literally a limb-for-limb comparison), using bone scaling and strength, safety (risk) factors, and ground reaction forces (the force generated when a foot pounds the ground). What was missing from these methods was a dynamic simulation of a moving dinosaur, one that encompassed all motions and forces of the animal during locomotion.

Such an approach would be our best option, because it required a complete set of modeling assumptions, and also because it was conceptually simple. The model required constructing a musculoskeletal system that made explicit estimates about the size and shape of the skeleton, body mass and mass distribution, together with muscle and tendon properties. These properties all have a considerable effect on locomotor performance, and these values could be inferred directly from our fossil evidence. Any unknown values needed to be clearly identified so that their importance to the final result could be assessed.

How could we validate such a model? Any model that attempts to predict the behavior of fossil species must be tested on extant animals. Sellers knew that a model attempting to predict the top speeds of a range of bipedal dinosaurs should also be tested using equivalent data from living bipeds. This is where we encountered a rather unusual problem. We found scant high-quality top-speed data for living species other than those actively involved in racing, such as humans, horses, and greyhounds. Although there are specific "speeds" quoted in the literature for ratites such as emu (*Dromaius novaehollandiae*) and ostrich (*Struthio camelus*), these are usually based on anecdotal observations. Such values need to be treated with caution. Chasing an emu or ostrich with a jeep across the savannah, and then gauging their speed from that of the vehicle, is not a very scientific approach.

Another major problem that Sellers and I have bemoaned is the fact I mentioned before: the anatomy, posture, and gait of bipedal dinosaurs were unique. Nothing walks or looks like a dinosaur today. Using any extant species as an analog is a useful starting point, but not an answer to understanding the locomotion of dinosaurs. We knew, however, that using the computational approach on extant species would validate the possiblity of generating a viable model that could be applied to extinct species.

The muscle model for the simulations was derived from existing published work. For the maximum running speeds we chose animals for which there were large amounts of existing published data. We chose humans, two extant avian theropods (birds), and five extinct theropods, all bipeds. These were very much two-dimensional models with a rigid trunk, and left and right thigh, shank, and composite foot. The limb segments were linked using three hinge joints per leg. The species that were chosen covered a reasonable size range and included the extant human (*Homo*), emu (*Dromaius*), ostrich (*Struthio*), and the extinct

Compsognathus, *Velociraptor*, *Dilophosaurus*, *Allosaurus*, and—a must for any dinosaur study—*Tyrannosaurus rex*.

The one parameter missing from all existing studies was accurate values for muscle mass in key locomotor muscles in the tail and limbs of dinosaurs. Sellers and I realized that we had to run the model first with existing data from other studies before we could use the priceless muscle-mass data from Dakota recovered from the CT scan data.

At the risk of becoming technical, I must point out that the gait of each extant and extinct animal was derived by using an algorithm that represented the gait cycle duration and the muscle activation levels at ten time periods through a gait cycle. Animals tend to work in the most efficient way to get from A to B for their specific body plan, geometry, the environment they are in, and what they are doing (hunting, breeding, or browsing). The computer searched for the most efficient way of getting from A to B for each model. The first few attempts were reminiscent of Monty Python's John Cleese and the Ministry of Silly Walks sketch, but each time the computer ran the model, it did so not from the start but from the adjusted place. After many thousands of cycles the computer came up with more efficient gaits for the model. It didn't so much learn from its mistakes as build upon them.

The fitness criterion applied to the algorithm was the greatest forward distance accomplished in a predetermined time. This meant that runs where the animal fell over scored very poorly and the runs with the greatest average speed would score the highest. The run population was 1,000, with up to 1,000 generations unless a stable maximum average forward velocity was accomplished earlier. This procedure was repeated at least five times until a good quality run was obtained for each of the extinct and extant animals. Runs were judged to be good quality when the animal did not fall over within the time limit and managed at least 15 meters of forward movement. The best run was then used by Sellers as the basis for a gait-morphing procedure, in which the best results for a previous run were used to generate the starting conditions for subsequent runs. This process was repeated at least 20 times to achieve the highest speed estimate for each species. Each individual species simulation ran in approximate real time, but at least 1,000,000 repeats were needed to generate the optimized running gait for each species. This repetition could take over a week for each species model.

Like watching a toddler take its first faltering steps, we watched the models slowly become steadier and much faster. By the end, all of the models generated high-quality running gaits that were stable over the entire simulation period. The top speeds achieved for each species were expected in some cases, but not in others. We found reasonably good correlation for the extant species between the speeds generated by the model and those cited in the literature.

We came to the conclusion that our multibody dynamic simulations using our new method appeared to provide reliable estimates for the maximum running speeds of the chosen extant animals. The multiple simulations, with small changes in both starting conditions and muscle activation patterns, produced highly consistent speed estimates. The many quoted values for running speed in the literature were based on observations made in less than ideal conditions, which may lead to considerable errors. Even for humans, the situation was not straightforward: while a 200-meter sprinter averages in excess of 10 meters per section, the peak speed reached can exceed 12 meters per second, but it has to be remembered that these are values for elite athletes who have considerably greater leg muscle mass than the average values used in our simulations. Research by others on more general female athletes from other sports give typical speeds of approximately 6 meters per second with short bursts of less than 8 meters per second. Our estimates were broadly in line with other biomechanical estimation techniques, which predict 18 meters per second for ostriches and 13 meters per second for emus. It was self-evident (and has been demonstrated in various models) that changes in muscle mass affected maximum speed, and this is a major source of uncertainty in such predictions. Ideally, what we needed was a dinosaur with a 3-D skin envelope bounding key muscle groups in the tail and legs.

Overall, the simulations illustrated how an animal could have moved given its physiological and morphological constraints, and perhaps also indicated probable movement patterns. Still we were some way off saying that this is how it must have moved.

SPRING IN THE TAIL?

The potential elastic properties of the backs of dinosaurs have fascinated Sellers and me for some time. The locomotor capabilities of quadrupedal vertebrates are considerably enhanced by the storage and recoil of elastic energy in the back tendons, providing energy recovery from step to step. However, elastic recoil in extant running bipeds, such as humans, is restricted to the legs and feet due to a vertical orientation of the torso. We decided to use a reverse-engineering approach to demonstrate that the unique body shape of bipedal dinosaurs enabled them to store energy in their horizontally held torso and tail. We had started this work just as the hadrosaur mummy was discovered, so we were still, to our frustration, too far away from data that we could use from Dakota's torso and tail. We decided to work on the predatory dinosaur *Allosaurus*, on which much has been published by other scientists. By tuning the resonant frequency of an elastic back spring in a computer simulation of *Allosaurus fragilis*, we were hoping to see if it would have an impact on locomotion speed and efficiency.

Having an internal, rigid, jointed skeleton is a crucial and defining adaptation of all vertebrates. This bodily system is moved by forces generated by muscles and acts as a series of levers to affect locomotion. Theoretical work on running, galloping, and hopping gaits showed that they could be modeled as a "mass on a spring." There is a theoretical zero cost for locomotion if the spring can store all the movement and gravitational energy lost on impact and then return it on takeoff. This use of elastic structures in the leg has been confirmed by experiments on humans, turkeys, and wallabies with approximately 40-50 percent of the energy recovered per step. Elastic structures in the spine seem to store energy during quadrupedal galloping, and this is apparently a mechanism for optimizing locomotion ability while reducing metabolic cost. The principle of elastic energy recoil is best displayed in the adaptation of the nuchal ligament, which allows long-necked animals, such as giraffes (*Giraffa camelopardalis*), to hold their heads high with minimal energy costs.

If elastic storage in limbs is of prime importance, other parts of the body, particularly the vertebral column, might also play important passive, elastic roles in locomotion. An old friend and mentor, Chris McGowan, had suggested that back tendons account for about 70 percent of the total elastic energy stored in the vertebral column, with muscles contributing 10 percent and the elasticity of the vertebrae themselves accounting for 20 percent. Could the same be said for dinosaurs? The vertebral column of mammals is quite different from that of reptiles, but given the similarities in the gait and posture with dinosaurs, their vertebral columns (as is probable in all vertebrates) most likely played an important role during locomotion. The ligaments and tendons that braced the backs of all dinosaurs would have played a crucial role in storage and release of energy during locomotion, as they do for mammals today, from shrews to elephants.

Predatory dinosaur skeletons display a typical form expected for a terrestrial cursorial animals (albeit a biped), but the posture of the limbs is not typical of reptiles, since the positioning of the hind limb allows an erect, or near-midline, posture and resultant gait (walking more like a bird). The erect posture, in which the plane of the legs is perpendicular to that of the torso, has the limbs slung under the body in a typically avian/mammalian posture. The vertebral column of *Allosaurus* was long, with a significant tail (40-plus vertebrae in many species) counterbalancing the torso, neck, and head. The effects of gravity on the vertebral column afforded many adaptations, resulting in a distinct specialization in the morphology (shape) and function of tail, body, and neck vertebrae. Sellers and I ran a simulation with and without elastic structures in the model's back, using the same software for the study of maximum running speeds. The results provided a complete picture of all the internal forces and strains within the model as well as energy flows and external forces. As a result we could investigate the interaction between these elements—an interaction that cannot be measured in physical experiments.

The theropod dinosaur *Allosaurus fragilis* was chosen as a representative bipedal dinosaur, since most agree that it was able to run and most of the required musculoskeletal parameters were available in the literature. However, no source was found for moments of inertia, so these were calculated by modeling the limbs as geometric shapes chosen to match the published lengths and centers of mass. Such abstractions led to moments of humor. By this time in our research we had started to refer to these simple models as "pointysaurus," reflecting the body shape of our virtual models.

Our results showed the maximum running speed clearly impacted the tail, body, and neck (collectively called the trunk) "springs" during the simulation of *Allosaurus*. It showed that a rigid trunk and a very stiff spring could approach the performance achieved, but a very compliant spring led to poorer performance. Stable running was impossible to achieve with very compliant springs, since they were unable to support the weight of the trunk segments. We were very happy with the results, but what would this mean for other dinosaurs?

Our simulations provided unequivocal support for the hypothesis that bipedal dinosaurs could have used their backs for storing elastic energy. It also showed that to make use of this storage the resonant frequency of the elastic system must be tuned to the stride frequency of the animal. All very elegant. This posed an interesting question regarding the function of the so-called ossified tendons (which may actually be ossified

own office in Manchester slowly fades into dusk as we speak; the modern communication world makes for a long working day!

Why create virtual dinosaurs? A computer simulation is much easier than attempting to work with actual specimens, because real fossil bones are often broken or distorted, and many of the pieces are missing. In the rare cases where the fossil bones are all available and sufficiently undistorted, trying to learn how they fitted together and how their joints operated by manipulating the original specimens is often quite impractical, owing to their fragility and enormous weight. Therefore, scientists create resin casts of the original specimens and manipulate those. But even then one either has to rig up a solid framework to support the casts (which then immobilizes them and makes subtle adjustments tedious or impractical), or to place them, say, literally in a sand pit, or to enlist enough helping hands to try to place them momentarily in space as they were envisioned to have been arranged when alive. Using these traditional methods, it is hard to learn about how even a relatively small region of the body such as a wrist or ankle functioned, since the involved bones need to be correctly positioned in space and to move relative to each other according to how the joints would have operated.

Stevens, in collaboration with paleontologist J. Michael Parrish, began to make virtual dinosaurs with software that he developed with his students at the University of Oregon. The first questions they examined were suggested by various researchers who have long wondered how the long necks of sauropod dinosaurs were posed, how flexible they were, and overall how these sauropods made a living. Starting from an engineering perspective, they needed to decide what to create in virtual space. The model had to capture enough of reality to be useful, but not so much as to be unwieldy. They decided to portray the necks as linked chains of bones, each with a limit on their ability to bend sideways and up and down. The bones could be schematically regarded as cylinders with ball-and-socket joints connecting the cylinders end-for-end. The model would also include the shapes of the small supplemental articular surfaces, which guide and limit the extent of deflection between each pair of neck bones. The complex shapes of the bones had more to do with muscle attachments, withstanding compression and bending loads, and weight saving. The simple representation of bones as cylinders plus little articular facets captured enough of the process to perform digital experiments on the undeflected pose and the flexibility of the necks of various sauropods.

Over roughly a decade of research, Stevens and Parrish found that sauropod necks shared a common design, of 15 (plus or minus a few) vertebrae, all arranged pretty much in a line, hanging out in front of the body as a straight extension of the backbone, not climbing up like the neck of a swan or giraffe. As Stevens's work became more internationally recognized, many museum directors must have held their head in their hands thinking about how they might have to remount their existing incorrect sauropod skeletons!

The modeling software Stevens developed was aptly dubbed DinoMorph, named for its goal of eventually capturing dinosaur morphology (shape), and also to permit "morphing" (the dynamic changing of shape, an effect familiar to viewers of computer graphics). Yet the early, primitive-yet-useful models in DinoMorph were sufficient only for asking some basic questions about necks. Asking more sophisticated questions about skeletal pose and flexibility would require far more sophisticated models of the individual bones. Fortunately, two trends in digital technology were maturing during this period. First, three-dimensional digitization by various means was becoming increasingly accessible, such as by laser scanner, CT scanner, and various devices that permit touching a probe to the object to be digitized. Second, digital sculpting techniques were becoming increasingly powerful, driven by the cinema and computer games industries. These two distinct means for inputting data into the computer would help capture the true form of dinosaur bones.

To digitize a fossilized dinosaur bone requires gathering a huge amount of data, which has to be simplified or pruned in order to create and view a whole skeleton of such "virtual bones." The data captures the morphology of that particular specimen, complete with any imperfections in the fossil such as twist, compression, and crushing distortions, cracks, and missing bits and chunks. If some repeating element such as a vertebra is completely missing, some stand-in must be provided. One practice is to replicate, with a bit of touch-up on the proportions and sizes, a similar bone. Soon, however, it becomes evident that the digitized models are at best faithful replicas of very imperfect real specimens.

The other technology that has been maturing along with scanning now comes in handy. Stevens has used sophisticated techniques for creating bone shapes provided by commercial software to create a generic bone shape for each of the various types of bones in a given dinosaur skeleton. Many of the bones vary subtly and smoothly along a skeleton, such as ribs and vertebrae. By creating a model for each type of bone, one that can be adjusted to match the actual data of a real fossil specimen, a nice blending of these two technologies can result. Digitization captures real data with its particularities of a specific bone, as well as the imperfections. Using that as a three-dimensional template for creating a highly accurate model permits capturing the important specifics of shape, to whatever level of detail is needed for the scientific task at hand. It also permits capturing the important morphology of the individual bones while avoiding the distortions and other imperfections of the raw data.

This is precisely what Stevens did for Dakota. The process of constructing the computer simulation of our dinosaur can be split into two phases. In phase one we needed to imput the required anatomical data into the computer. The first step in this process was giving Stevens a reference hadrosaur skeleton. This was achieved by laser-scanning a composite cast of *Edmontosaurus* held at the Black Hills Institute. Using a cast is helpful because the major deformations in the fossils had already been largely corrected, and the casts themselves were made of lightweight foam, permitting easy manipulation of the bones.

Sellers and I used a handheld Polhemus laser scanner, which allowed the laser wand to be moved over the bones. The bones themselves could also be moved to allow more convenient scanning, which was essential for complex shapes such as the vertebrae. The Polhemus system uses a magnetic field to measure the orientation and position of both the bone and the scanning wand and performs the necessary geometrical calculations to produce properly aligned scans of the whole bone. This procedure was performed under the watchful eye of Stevens, who would sit alongside strapping vertebrae together with rubber bands, often scratching his head while in deep thought. This is the fun part of working with dinosaurs, collaborating with such remarkable scientists.

These scans were then imported into the program at the University of Oregon, where any remaining distortions owing to the fossilization or scanning process could be corrected. Each bone was then painstakingly assembled into a complete, posable skeleton using the Dinomorph software. The modeled bones could now fit together without distortions, and repeating bones, such as the many vertebrae or ribs, could be created from a relatively small number of prototypes along the length of those series of bones.

By manipulating the separation and angles between bones on either side of each joint, Stevens was able to replicate a complete mount of the *Edmontosaurus*. Because it exists in virtual space, there is no problem with adjusting any separation or joint angle, throughout the skeleton. The range of motion of various joints needed for locomotion was then estimated, based on what is generally understood about the limits on joint flexibility in various modern animals. Estimated neutral standing poses (both bipedal and quadrupedal) and ranges of motion for the major limb joints were provided to Sellers for use in his

GaitSym software. A distribution of body mass was also estimated, by associating simplified slices of body volume (times the density associated with that slice) with segments of the body. The mathematical center of mass, or balance point, could then be estimated and also provided to Sellers.

DINOMUMMY GAIT SIMULATION

Traditionally, reconstructing the locomotion of extinct animals has been a largely artistic endeavor in which highly skilled animators use their knowledge of animal locomotion to create walking or running gaits that "look right." Sometimes experts are brought in to express their opinions about the verisimilitude of the final result based on their knowledge of dinosaur biology and preserved trackways. There is, however, generally little or no biomechanical or physiological input, and until recently this would only have helped make the expert a little more informed. Recent advances in computer technology have transformed this situation. The computer can now generate locomotion, as we discussed earlier, based on all the information we can amass about the anatomy and physiology of the fossil animal. These simulations are mechanically consistent and require no prior specification of the expected gait. They are objective reconstructions based on best current scientific knowledge.

We have not yet reached a point where this technology is a panacea. The current level of computer technology is such that the simulation for even a simple model takes weeks to evaluate on a supercomputer. This is science, not animation. These strictures limit the models to two-legged dinosaurs and mean that we have to simplify the anatomy so that only a few large leg muscles are represented. Fortunately, computers are getting better and cheaper all the time, and more detailed models, including quadrupeds, are currently being developed; they should be available in the next few years. Another limitation is the lack of information about some of the key anatomical parameters that the models need. In particular, even after decades of work, it is still very hard to estimate the bulk of dinosaurs, and this is obviously essential if we are to work out how they moved. A large-bodied animal on spindly legs is going to be much slower than a sleek animal on well-muscled legs.

This is why the dinosaur mummy is such an important fossil. Dakota has a largely complete skeleton, so we can estimate the stature of the animal accurately, and by having a preserved skin over much of the body, we can for the first time reliably estimate the bulk of the animal. The skin allows us to calculate the surface area of the various parts of the body (the torso, tail, thigh, calf, and foot regions), and when Sellers and I reconstructed the animal, we could calculate the volumes of these parts. We know the density and tissue components of these parts of the body from studies of other related animals, so we can calculate both an accurate body mass and also sizes of the muscles. This removes the most important source of uncertainty from our simulations, leading to the most accurate set of estimations to date.

The skeleton and body outlines built by Stevens were transferred across the Atlantic to Manchester, where they could be used in our gait simulation. A simplified set of muscles was attached to the skeleton using the visible bony landmarks on the bones to identify attachment points. The presence of the skin envelope allowed accurate estimation of muscle size. The body outline was also used to calculate the mass parameters associated with each part of the body that is needed for mechanical analysis.

Joints were created in the simulation representing the anatomical joints from the skeleton, and contact points were generated to allow the feet of the model to interact with the ground in the virtual gait simulator. The simulator has been written in-house, by Sellers, to take advantage of multiple parallel processors in a super-computer. It models Newton's laws of motion and was able to calculate the movements of the parts of the dinosaur's skeleton under the influence of muscle forces, gravity, and the forces transmitted through the feet.

The model of our hadrosaur was now ready to learn how to walk or run. This process is very slow, since the computer needs to find out how to activate the muscles to generate a stable gait. Obviously, there are an enormously large number of possible activation patterns, and there are a large number of possible gaits. Much like humans, bipedal dinosaurs would have been able to use a full range of walking, skipping, hopping, jumping, and running gaits. We needed to find a way of choosing both the gait and the activation pattern, and to do this we used a simple rule: Go as far as possible in five seconds. Sometimes the simple questions in science are the best! Such a condition should find the fastest physiologically possible gait for the animal while allowing the computer to use any gait pattern it likes.

When we wanted to find slow, efficient gaits we used the rule: Go as far as possible for the stipulated amount of energy. Still, the computational problem is extremely difficult, and so we have borrowed a search technique from artificial intelligence called the genetic algorithm. As we discussed earlier, this algorithm works by starting off with 1,000 random activation patterns of movement. Once these are all evaluated, the vast majority prove to be no use at all. The simulation falls over very quickly and grinds its virtual face in the dirt. However, a few simulations progress forward a little bit, and these ones are used as a starting point to create another 1,000 random activation patterns. If we are lucky, some of these new patterns will work a little bit better than the previous ones, and these can then be used as a new starting point for another 1,000 random activation patterns.

In this way, by selecting the best solutions each time and using them to generate the next batch of activation patterns, we see a slow improvement, and we end up with an activation pattern that produces high-quality gait. In general 1,000 repeats of the whole process will produce something quite good—as in 1,000,000 runs of the simulation—but since there is a random element in this process, we need to repeat it a number of times to be sure we are getting the best result. All told, the process takes a huge amount of time.

The simulation with an optimized activation pattern produces a fully animated, 3-D description of the animal's movement. Unlike our "pointysaurus" model, we now had a rather attractive 3-D musculoskeletal model of our hadrosaur. We used this movement to estimate the locomotor abilities of the animal, such as its top speed and the amount of energy it used traveling. We could also produce high-quality animated sequences by exporting the 3-D geometry frame by frame and using standard computer-graphic techniques. These can then be displayed on the Internet or used in TV programs or as still images.

Currently, we have a functioning simulator and high-quality animations for the generic hadrosaur. We are still waiting for the final analysis of the CT data to obtain the specific dimensions of the dinosaur mummy so that we can produce the high-accuracy simulation. In particular, we have been adding additional muscles to the simple muscle so that the legs of the simulation more accurately represent the anatomy of the animal. These include muscles that cross more than one joint in the animal, since this is commonly the case. Alas, until we get the final CT data from Anders at Boeing/NASA, we have to refrain from lighting the victory cigar!

EPILOGUE

AS THE PAST 18 MONTHS HAVE SHOWN, science has a habit of not always going according to plan. The methods that we are now applying in the study of this remarkable fossil in many cases were not planned at the beginning of the project. As we discovered new secrets about our prehistoric parcel, we employed novel techniques to tease out more truths about life in the era of dinosaurs. Some are so innovative that we are still sifting through the data and discovering how to interpret the findings.

The excavation itself provided a number of lessons. The physical access to and from the site was a logistical nightmare; moving tons of supplies (plaster, water, steel frames, digging equipment, people, etc.) to and from a very isolated locale made for long and often tiring days. However, the accommodation back at the bunkhouse every night was akin to luxury when compared with camping in the Badlands. We were as prepared as we could be for health and safety at the site, but it is hard to rule out every potential mishap, as Steven Cohen discovered when the scorpion climbed up his shirt, and we were pleased that we did not add to the four percent fatality figure for rattlesnake bites.

The steel frame that Tyler's brother welded onto the base of the body block proved to be one of the great successes of the dig. It allowed the lifting and transportation of the block without causing any harm to the fossil inside. The field jackets, although cumbersome, also served their purpose. Tyler was right to be overcautious with the application of plaster and burlap. The fact that no glue or resin was used on the block also assured that samples taken for biomolecule analysis had as little contamination as possible. In the future I will also bring an organic-sampling field kit, so that if any material is accidentally exposed during excavation, samples will be taken immediately and stored in inert glass vials. Any such samples might also be mapped in LiDAR digital outcrop models to add to the special map of sediment chemistry around a fossil.

We started our sediment study by collecting field samples, making sure that we recorded every shred of data. The samples would also have to provide potential data for future studies or techniques that might test our own findings. This process would hopefully future-proof our study. In every case, we took large samples so that only a fraction of any one sample was processed at a given time, leaving reference material for future studies to either duplicate results or apply new techniques. The use of 21st-century surveying techniques made the subsequent analysis easy and accurate.

The LiDAR study provided a perfect 3-D digital outcrop model within which we could place sedimentary data, and it also allowed us to model large-scale features, such as the river channel cutting alongside the excavation site. The presence of the channel so close to the site explained much of Macquaker's observations of the sediment chemistry. The river would have kept plenty of water pushing into the sedimentary deposits on either side of the channel, assuring that the mineralogical soup had a constant supply of organic-rich waters. The organic material in the water would almost certainly have come from the large quantities of plant material deposited within the channel and floodplain deposits.

Macquaker, Gawthorpe, Hodgetts, Taylor, and Marshall have pieced together the paradoxical puzzle resulting from Dakota's unique preservation. The combination of immature, reactive minerals present in the sediments; humic acids (a function of plant debris and a large volume of water); and the chemistry of death and decay (tissue breakdown and microbial activity) provided the ideal conditions for the rapid mineralization of the animal's soft tissues. Dakota was a dinosaur buried in the right place, at the right time, in the right sediment, with the perfect chemistry and a suitable community of microbes to ensure its preservation. Can its state of burial help us find other such dinosaurs? The fact that we can now characterize one ideal environment that successfully preserved a mummified dinosaur means that we can hunt for similar windows in the geological record.

A good place to start hunting is along the edge of other large fossil channel deposits. However, the channel itself is not sufficient; the mix of the ingredients mentioned previously is also necessary. I have a feeling that the Sternbergs had made similar observations of the conditions at their 1908 mummy site, since they thereafter made a habit of finding such fossils. The first mummy was more luck than judgment; I think subsequent ones were more judgment than luck.

Although the Sternbergs had prepared their first mummy at the excavation site—a fact I find incredible—we were not so brave. Besides, we had a great deal of data that we wished to recover for testing, and prepping the mummy on-site would have been disastrous for those purposes. The many folks who have spent thousands of hours picking away grains of sand from the delicate fossil tissue structures on Dakota have had the patience of Job. This process will continue for a few more years, as painstaking scientific caution cannot be hurried. We are hopeful that the CT scan eventually provides a road map to aid the process. We have certainly learned many new skills and methods from Dakota's preparation, most from the patient hand of Stephen Begin in studying its arm.

The CT scanning has taught me one valuable lesson: When you are trying to push the envelope in physics, leave yourself plenty of time. The CT images continue to elude us, but not for lack of trying. Jeff Anders continues to battle with the 8,000-pound body block at the Boeing/NASA facility. I received an e-mail from him recently, saying that he was going to rebuild one of the "harder" x-ray linear accelerators to complete the project. Knowing now what I do about Anders, I understand how NASA managed to put a man on the moon!

As for the mystery of the *Borealosuchus* crocodile that is so intimately associated with our mummy, that tale will have to remain untold until we receive the CT data or prep down to its fossil remains. The articulated hand and forearm remain a tantalizing "wave" at the paleontologists who have seen this remarkable fossil. Who knows—if one crocodile came to feed on poor Dakota's remains, there may be more!

The biggest bonus of the whole project has to be the astonishing amount of detail preserved in Dakota's fossil skin. When I first saw what Tyler had unearthed, I have to admit that I thought I was viewing skin impressions. Our subsequent analysis shows that the skin had a 3-D component to it, with scales beautifully raised in relief on many parts of the body and limbs. That depth of the skin attracted Roy Wogelius to the project, and he in turn introduced me to the analytical world of the geochemist. Wogelius continues to head the geochemical and biomolecule research with a team of committed scientists.

What has heartened me above all else is how many folks have given their time to help unlock the secrets of the fossil. The samples that Tyler supplied of fossil skin and keratinous sheath from the mummy have opened up a whole new direction of research for the Manchester team. We are encouraged by the preliminary results, and all team members are sure we have recovered original biomolecules from Dakota.

The recovery of potential amino acids and other breakdown products from Dakota's original structural proteins, combined with the imaging and recognition of soft-tissue structures, has surprised everyone involved with the project. Five different labs in two different universities have confirmed our results. We are currently waiting for results from a sixth lab at a third university—nothing like being absolutely sure! Some might even accuse us of being overcautious, but that is an accusation that both Wogelius and I are happy to live with. The hunt for an intact protein molecule continues. All on the team feel there is an outside chance that one might have survived in the vast body block. The fact that the CT work has slowed much of the project is possibly fortunate, as we now have a better idea of what to look for when sampling potential sites for organic material. The body block might provide an organic mine for the next few years.

The locomotion studies undertaken by Bill Sellers, Kent Stevens, and me continue apace. The skillful modeling of Dakota's skeleton by Stevens will evolve as more CT data is applied to further define the scale and geometry of the dinosaur's skeleton. Stevens continues his dinosaur locomotion modeling in Oregon, while Sellers and I refine the preliminary model back in Manchester. The information that we have gleaned about the form and function of the skeletal system of dinosaurs in the course of this study continues to spring surprises. Once the skin envelope data from Dakota's tail is wrapped around the skeletal computer model, with the muscle mass constrained, we will be that much closer to understanding what it was really like to be walking with dinosaurs.

One emerging component of the project that I have not discussed is the exploration of hadrosaur respiratory biology. Dinosaurs, if they were like their bird descendants, would not have had a diaphragm to assist inhalation. The biology and physiology of dinosaur respiration have thus received a great deal of attention, and we hope that Dakota will help resolve some of the many questions that have been raised. A team member you have not met is respiratory biologist Jonathan R. Codd from the University of Manchester. He has been working for many years on how birds breathe. Codd and I have recently joined forces to study the breathing mechanics of predatory dinosaurs, specifically the maniraptoran theropods that gave rise to birds. We are also keen to resolve the function of distinct cartilaginous plates present on the rib cage of many ornithischian dinosaurs, and possibly present in Dakota as well. These plates might have functioned as respiratory "rubber bands," pulling the rib cage of the dinosaur back into position during exhalation. However, until the preparation of Dakota's rib cage is complete or the CT scan reconstructed, we will be unable to proceed with this area of the project.

One fact is certain: We have been able to review much of what we thought a hadrosaur dinosaur looked like. The Sternberg mummies had already provided large amounts of data on the form, geometry, and location of areas of skin on the body of hadrosaurs, but Dakota has helped to fill in many of the gaps, especially the tail, arm, and hand. The continued studies of the skin envelope in coming months and years will help piece together the hide of a hadrosaur. When we have a complete skin envelope, we will be able to fill it with flesh and bone to accurately measure the volume of a dinosaur for the first time. Once we have the volume, we can use data on the density of reptiles and birds to generate a mass for our dinosaur. We will soon have the most accurate weigh-in of a large dinosaur in history!

You may recall my fanciful reconstruction at the beginning of this book, when we followed Dakota on his last hours alive on Earth. Before I review that scene, I must stop myself from using "him" or "his" until the prep or CT work yields the sex of Dakota. The cloacae—the openings for the animal's genitals—may be preserved, especially given the complete nature of the skin envelope on the tail. Thankfully, whether the dinosaur turns out to be male or female, the name will be transferrable.

As for the opening scene of this book, the large river in which Dakota came to grief definitely existed on the Hell Creek plain. The astounding level of preservation supports the swift burial of our animal, given that scavengers and microbes were suppressed long enough to allow the rapid mineralization of soft tissue. Subsequent floods almost certainly buried the carcass soon after the *Borealosuchus* ate its fill or drifted up against Dakota. The cause of death remains a mystery, for the time being. However, the impending CT results and future preparation might just reveal it.

DINOSAUR NEEDS A HOME

Much work remains to be done on the project, not least the continuation of the preparation of the fossil remains. The initial funding from the National Geographic Society and the University of Manchester, coupled with the generous support from Boeing Corporation and NASA, has made so much possible, but the project requires additional funds to continue the labwork and science.

The Marmarth Research Foundation is seeking funds to build a museum that would display the fossil remains of Dakota and the many other fossils from the Hell Creek Formation that its staff and volunteers have lovingly excavated and prepared. Development of the museum can happen—the Royal Tyrell Museum in Drumheller, Canada, is tangible proof of this—and convincing a sponsor is what we hope public awareness of Dakota will achieve. Such a rare dinosaur deserves a home in Marmarth, where it would receive the care and curatorial expertise required to conserve this remarkable fossil.

In the meantime the CT, organics, locomotion, sedimentology, and scientific exploration of Dakota will continue for many years. Samples recently sent to the University of Bristol Organic Analysis Laboratory may shed supporting light on many of the soft-tissue structures, and a recent conversation with Mary Schweitzer at the Symposium of Vertebrate Paleontology in 2007 has ensured trans-Atlantic collaboration on the same topic. The future for Dakota is bright.

When I was first contacted by Tyler Lyson, after that fateful conversation with Emma Schachner, I would not have believed the complex journey that Dakota has taken us on. Usually after such a grueling excavation and laborious scientific undertaking I am shattered to the point where I swear that I never want to dig another dinosaur again. However, both Tyler and I feel different about Dakota. Here is a fossil that has captured the hearts and imagination of all who have looked upon the immortal remains.

ACKNOWLEDGMENTS

This book has been a sprint to the finish line. The combined research, field, and teaching programs conspired to make the completion of this book an interesting experience, which has been made more comfortable by the support of many people. First, I would like to thank Joanne, Alice, and Kate for putting up with my hours of lock-away while writing this book. I would especially like to thank Tyler, Rance, and Molly Lyson and the wonderful folks at the Marmarth Research Foundation who made so much in this book possible.

I also offer deep thanks to the National Geographic Society for grants from the Committee for Research and Exploration and the Expeditions Council, which supported much of the field and laboratory work undertaken in the study of Dakota. Special thanks are due to Rebecca Martin and Terry D. Garcia of the Expeditions Council and Missions Programs, respectively, for their support and helpful council. Thanks must also be given to the National Geographic Television and Film team, who included Jenny Kubo, French Horwitz, Sarah Meyer, Chad Cohen, and Maryanne Culpepper, who often helped hugely in organizing access to key facilities for research to be undertaken. My thanks to editor Garrett Brown and publisher Kevin Mulroy in the book division for their patient support.

I would especially like to thank Boeing, NASA, and Pratt and Whitney for access to their CT facilities and funding to acquire the vast CT data sets. I must thank Jeff "Determined" Anders above all at the CT facility and apologize to his wife and family for the hours he has had to work on this project! Thanks must also go to HESCO, who kindly donated the use of their 9-MeV linear accelerator to "upgrade" the CT capabilities of the project.

As the project developed, many key individuals offered their time and facilities to make so much of the science possible, especially the Wolfson Molecular Imaging Centre at the University of Manchester. Many people offered comments on the manuscript as it was rapidly developing; for this I thank Dr. David West Reynolds, Professor John Prag, Dr. Duncan McIlroy, Professor Martin Brasier, Dr. Roy Wogelius, Professor Jim Marshall, Dr. Andy Gize, Professor Adam McMahon, Dr. Emrys Jones, Professor Piotr Bienkowski, Professor Kent Stevens, Dr. Andrian Schofield, and Dr. Joe Macquaker for their insightful comments and content.

The members of the mummy team who have contributed so much to this project you have met in the pages of this book. I cannot thank the team members enough for their time, commitment, and determination. I am also deeply indebted to Emma Schachner for putting me in touch with Tyler. Finally, I thank Professor Chris McGowan, who has given me both the inspiration and confidence to write my first book. You're right, Chris: Once you start writing, it's hard to stop!

ABOUT THE WRITER

Dr. Phil Manning is a paleontologist, fossil hunter, and writer who has also contributed to many natural-history television documentaries. He graduated in 1988 from Nene College (now the University of Northampton) in the United Kingdom, where he studied meteorology, applied climatology, and Earth sciences. He went on to get his master's degree at the University of Manchester on eurypterid paleoecology, and then his Ph.D. at the University of Sheffield on dinosaur tracks and locomotion. Manning has concentrated his academic work on understanding the complex relationships among sediments, dinosaur limb morphology, and kinematics to unravel the secrets of locomotion locked with the 3-D fossil tracks of dinosaurs.

Manning has been an active member and council member of the Paleontological Association London and was recently appointed as Vice Chair of the International Year of Planet Earth outreach program. He has worked in museums on the Isle of Wight, Clitheroe, York, and Manchester and has held several curatorial positions. He has also taught vertebrate paleontology and evolution at the universities of Liverpool and Manchester. At Manchester, Manning heads the vertebrate palaeontology research group. His writing is both broad and diverse; he has published papers on dinosaur tracks, theropod biomechanics, arthropod paleobiology, vertebrate locomotion, and the evolution of flight in birds.

Each year Manning spends a great deal of time in the field collecting data for his ongoing research but also devotes a large amount of time to delivering public lectures on paleontology and the evolution of life on Earth. He lives in Manchester, with his wife, Joanne, and two daughters, Alice and Kate.

ABOUT THE COVER ILLUSTRATOR

Julius T. Csotonyi is a self-taught Canadian artist specializing in digital and traditional rendering of prehistoric life. Since 2005, he has worked closely with paleontologists to create illustrations for more than two dozen books and museums. Following on his graduate studies in ecology, Csotonyi has also published several scientific papers on diverse life-forms—from mutualistic desert plants and animals to bacteria from the deep sea.

PHOTOGRAPHY CREDITS

All interior photographs courtesy Phillip Manning, unless otherwise noted: Cover, Julius Csotonyi; 2-3, Julius Csotonyi; 189, DK Limited/CORBIS; 190-191 (UP), Brooks Walker; 195, Jenny Kubo, NGS; 199, Jenny Kubo, NGS; 201, Courtesy Dave Hodgetts; 204, Jenny Kubo, NGS; 221, Pratt & Whitney; 320, Julius Csotonyi.

Look for this amazing story on the small screen in *Dino Autopsy* on the National Geographic Channel.

For more information on other National Geographic books and DVDs about dinosaurs, visit www.shopng.com/dinos.

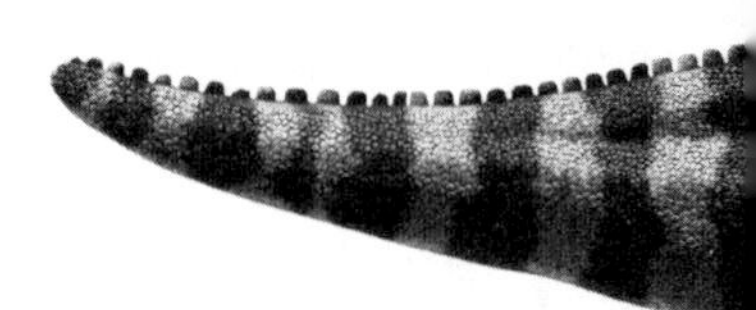